MATLAB for Electrical Engineers and Technologists

MATLAB tutorial with practical electrical examples

Stephen P. Tubbs, P.E.
formerly of the
Pennsylvania State University,
currently an
industrial consultant

NOTICE TO THE READER

The author does not warrant or guarantee any of the products, equipment, or software described herein or accept liability for any damages resulting from their use.

The reader is warned that electricity and the construction of electrical equipment are dangerous. It is the responsibility of the reader to use common sense and safe electrical and mechanical practices.

Printed in the United States of America and United Kingdom.

ISBN 978-0-9819753-2-0

AMD Athlon 64, AMD Opteron, and *AMD Sempron* are trademarks owned by American Micro Devices, Inc.

Debian is a trademark of Software in the Public Interest, Inc.

Intel Atom, Intel Celeron, Intel Core, Intel Core 2 Duo, Intel Pentium 4, and *Intel Xeon* are trademarks owned by Intel Corporation.

GNU Octave is copyrighted by John W. Eaton.

Linux is a trademark of Linus Torvalds, the original author of the Linux kernel.

Mac and *Mac OS X* are trademarks owned by Apple Inc.

Maple is a trademark owned by Maplesoft, Inc.

Mathcad is a trademark owned by Parametric Technology Corporation (PTC).

MATLAB, Simulink, SimPowerSystems, and SimElectronics are trademarks owned by MathWorks, Inc.

Mathematica is a trademark owned by Wolfram Research, Inc.

Microsoft Excel, Microsoft Word, Windows, Windows Server, Windows Vista, and *Windows XP* are trademarks owned by the *Microsoft* Corporation.

OpenSUSE is a trademark of Novell, Inc.

Red Hat Enterprise Linux v.4 is by Red Hat, Inc.

Scilab is a trademark owned by INRIA.

Solaris is a trademark of Sun Microsystems, Inc.

SPARC and *ultraSPARC* are trademarks of SPARC International, Inc.,

Ubuntu is a trademark of Canonical Ltd.

CONTENTS

PAGE

INTRODUCTION--- vii

1.0 WHAT IS MATLAB?-- 1

2.0 MATLAB AND POPULAR ALTERNATIVE PROGRAMS-------------------------- 2

2.1 MATLAB--- 2

2.1.1 MATLAB 30-DAY FREE TRIAL DOWNLOAD----------------------- 2

2.1.2 MATLAB AND SIMULINK STUDENT EDITION-------------------- 2

2.1.3 MATLAB INDIVIDUAL LICENSE PROFESSIONAL
VERSION--- 4

2.1.4 MULTIPLE USER VERSIONS-------------------------------------- 5

2.2 ARE THERE MATLAB ADD-ON PROGRAMS RESIDING
IN YOUR COMPUTER?-- 5

2.3 MATLAB TRAINING AND TUTORIALS------------------------------------- 6

2.4 HOW POPULAR IS MATLAB?-- 6

2.5 MATLAB CLONES-- 6

2.6 HOW DOES MATLAB COMPARE WITH SIMILAR PROGRAMS?------------ 7

3.0 EXAMPLE PROBLEMS-- 8

3.1 INTERACTIVE MODE (USING MATLAB AS A CALCULATOR)------------ 9

3.2 STEADY-STATE ELECTRICAL CIRCUITS------------------------------------ 11

3.2.1 SIMPLE DC CIRCUIT-- 11

3.2.2 DC MESH CIRCUIT--- 18

3.2.2.1 DC Mesh Circuit Solution using Determinants------------------ 20

3.2.2.2 DC Mesh Circuit Solution using MATLAB's Matrix
Left Division--- 23

3.2.2.3 DC Mesh Circuit Solution using MATLAB's
Matrix Inversion--- 26

3.2.3 SIMPLE AC PHASOR CIRCUIT------------------------------------- 29

3.2.4 AC PHASOR MESH CIRCUIT-------------------------------------- 32

3.2.5 AC INDUCTION MOTOR ANALYSIS-------------------------------- 37

3.2.5.1 Solution using a Try and Test Program------------------------ 38

3.2.5.2 Solution using the Optimization Toolbox "fsolve"
Subroutine--- 44

3.2.6 AC INDUCTION MOTOR 2-D PLOT OF EFFICIENCY
VERSUS SLIP--- 52

3.2.7 AC INDUCTION MOTOR 3-D PLOT OF EFFICIENCY
VERSUS SUPPLY FREQUENCY AND SPEED----------------------- 57

3.2.8 SIMPLE STATIC DC DIODE CIRCUIT-------------------------------- 64

3.2.9 SIMPLE STEADY-STATE AC DIODE CIRCUIT--------------------- 67

3.2.10 INDUCTOR CHARACTERISTICS----------------------------------- 72

3.3 TRANSIENTS IN ELECTRICAL CIRCUITS------------------------------- 78

3.3.1 RL CIRCUIT, FIRST ORDER DIFFERENTIAL EQUATION--------- 78

3.3.2 RLC CIRCUIT, SECOND ORDER DIFFERENTIAL
EQUATION--- 81

3.3.3 AC INDUCTION MOTOR DRIVING A RECIPROCATING
PUMP, SYSTEM OF FIRST ORDER DIFFERENTIAL
EQUATIONS--- 85

3.4 MISCELLANEOUS EXAMPLES--- 90

    3.4.1 STATISTICAL ANALYSIS OF RESISTOR RESISTANCES--------- 90

    3.4.2 APPROXIMATING RESISTANCE VERSUS TEMPERATURE
          DATA WITH A POLYNOMIAL--- 94

    3.4.3 DETERMINING THE FREQUENCY CONTENT OF A
          WAVEFORM WITH THE DISCRETE FOURIER TRANSFORM--- 97

    3.4.4 FOURIER SERIES ANALYSIS--- 104

    3.4.5 SIMULINK EXAMPLE-- 112

4.0 REFERENCES--- **124**

5.0 APPENDIX-- **125**

    5.1 NUMERIC PRECISION-- 125

    5.2 TIPS-- 125

    5.3 MATLAB OPERATORS--- 125

INTRODUCTION

According to its website, "MATLAB is a high-level technical computing language and interactive environment for algorithm development, data visualization, data analysis, and numeric computation." The website also states, "You can use MATLAB in a wide range of applications, including signal and image processing, communications, control design, test and measurement, financial modeling and analysis, and computational biology. Add-on toolboxes (collections of special-purpose MATLAB functions, available separately) extend the MATLAB environment to solve particular classes of problems in these application areas."

Its website states, "Over 1,000,000 engineers and scientists in more than 100 countries, on all seven continents, use MATLAB and Simulink. These products have become fundamental tools for work at the world's most innovative technology companies, government research labs, financial institutions, and at more than 5,000 universities." Further evidence of its popularity is the large number of other numerical analysis programs that are MATLAB-compatible, MATLAB-like, or interface-able to MATLAB.

The first purpose of this book is to quickly teach an electrical engineer or technologist how to use MATLAB. The reader learns by example. Complete keystroke-to-keystroke details are provided for problem solution and documentation. Most of this book's examples demonstrate MATLAB's abilities as a stand-alone programming language for performing numeric electrical computations. Also, two MathWorks add-on programs are demonstrated, the Optimization Toolbox and Simulink.

The second purpose of this book is to demonstrate MATLAB solutions of practical electrical problems.

The simplest and most basic uses of MATLAB are in the first examples. Later examples demonstrate more complex capabilities. The reader could use the examples' solutions as starting models for his own programs.

It is assumed that the reader has an analytical electrical background of the sort that would be gained in a university electrical engineering or technology program.

MATLAB is available in a free 30 day Demonstration version. Its key features can be learned in 30 days.

Stephen Tubbs, P.E.

1.0 WHAT IS MATLAB?

MATLAB is an interpretive programming language and an operating environment for many add-on programs.

MATLAB programs are often called M-files or scripts.

MATLAB is not a compiler language. It executes commands line by line rather than compiling all the lines and making an executable code before executing its program. MATLAB can be used in an interactive mode or program mode. In the interactive mode, arithmetic or mathematical statements are typed into MATLAB and it immediately calculates and produces an answer. In the program mode, MATLAB appears similar to a compiler language. In this mode, a program is typed on a text processor and then loaded into MATLAB. Then it runs the program.

MathWorks, Inc. created MATLAB. MAT is short for **mat**rix and LAB is short for **lab**oratory. It is especially good at matrix manipulations. It also plots data and functions and interfaces with other programming languages. MathWorks and other companies make many specialized add-on programs that use Matlab as an operating environment.

MATLAB was created in the late 1970's. In 1984 The MathWorks, Inc. was founded and it produced a C based version of MATLAB. Development has continued, MathWorks now creates two core products, MATLAB version 7.9 and Simulink version 7.4. MATLAB can do useful numeric computations alone. Simulink requires integration with MATLAB. There are MATLAB and Simulink families of products.

The MATLAB product family includes 47 optional add-on programs called "toolboxes". These "toolboxes" cover topics such as parallel and distributed computing, math and optimization, statistics and data analysis, control system design and analysis, signal processing and communications, image processing, test and measurement, computational biology, computational finance, application deployment targets, and database connectivity and reporting,

The Simulink product family includes 34 optional add-on programs. These programs cover topics such as report generation, fixed-point modeling, event-based modeling, physical modeling, simulation graphics, control system design and analysis, signal processing and communications, code generation, rapid prototyping and HIL simulation, verification, validation, and testing.

This book was written using MATLAB Version 7.5.

2.0 MATLAB AND POPULAR ALTERNATIVE PROGRAMS

2.1 **MATLAB**

Information on all versions can be found through The MathWorks, Inc. website, http://www.mathworks.com. The MathWorks, Inc. is located at 3 Apple Hill Drive, Natick, MA. They can be contacted by phone at 508-647-7000 or fax at 508-647-7001.

Prices given here were in effect at the time of writing, check with The MathWorks for current prices and packages.

2.1.1 MATLAB 30-DAY FREE TRIAL DOWNLOAD

The trial version is the same as the full Professional Version of MATLAB except that it expires 30 days after it is installed.

Specific instructions on the downloading, installation, and start-up of the MATLAB program can be found on the MathWorks website.

2.1.2 MATLAB AND SIMULINK STUDENT VERSION

MathWorks has created a student version software bundle of 32-bit MATLAB, Simulink, Symbolic Math Toolbox, Control System Toolbox, Signal Processing Toolbox, Signal Processing Blockset, Statistics Toolbox, Optimization Toolbox, and Image Processing Toolbox. The current release is 2009a. Its list price is about $100. Additional student version add-on programs may also be purchased for $59.00 per program.

The MATLAB Student Version provides all the features and capabilities of the professional version of MATLAB software, with no limitations. There are only a few small differences between the Student Version interface and the professional version of MATLAB, the prompt is EDU>>, printouts contain the footer "Student Version of MATLAB", and the SimMechanics software has a 20 body limitation. Also, the add-on programs intended for the Student Version are not compatible with the Professional Version and visa versa.

The Student Version is only for students who are currently enrolled in a degree-granting college, university, high school, or continuing education program. All students must submit proof of student status. The programs are to be used for course work. Their use in professional research at universities or in commercial enterprises is strictly prohibited.

Instructors may apply for an evaluation copy of the MATLAB and Simulink Student Version to help them prepare for their classes.

SYSTEM REQUIREMENTS FOR THE STUDENT VERSION

WINDOWS OPERATING SYSTEM
Hardware Requirements
Intel Pentium 4, Intel Celeron, Intel Xeon, Intel Core, AMD Athlon 64, AMD Opteron, or AMD Sempron.
512 MB of RAM (At least 1024 MB recommended).
660 MB of hard disk space (without any MATLAB add-on programs).
16-, 24-, or 32-bit OpenGL capable graphics adapter.
Software Requirements
Windows XP Service Pack 2 or Service Pack 3 or Windows Vista Service Pack 1.
Microsoft .Word 2002, 2003, or 2007 to run MATLAB Notebook.
Microsoft Excel 2002, 2003, or 2007 to run Spreadsheet Link EX.

MAC OPERATING SYSTEM
Hardware Requirements
All Intel-based Macs.
512 MB of RAM (At least 1024 MB recommended).
360 MB of hard disk space (without any MATLAB add-on programs).
Software Requirements
Mac OS X 10.5.5 and above.

LINUX OPERATING SYSTEM
Hardware Requirements
Intel Pentium 4, Intel Celeron, Intel Xeon, Intel Core, AMD Athlon 64, AMD Opteron, or AMD Sempron.
512 MB of RAM (At least 1024 MB recommended).
500 MB of hard disk space (without any MATLAB add-on programs).
16-, 24-, or 32-bit OpenGL capable graphics adapter.
Software Requirements
Red Hat Enterprise Linux v.4 and above, Fedora Core 4 and above, Debian 4.0 and above, or Ubuntu 8 and above.

2.1.3 MATLAB INDIVIDUAL LICENSE PROFESSIONAL VERSION

MATLAB is available in both 32-bit and 64-bit.

The Professional Version of MATLAB is purchased by itself. Simulink and other add-on programs would be purchased in addition to it. Some of the current list prices are: MATLAB $2000, Simulink $3000, Symbolic Math Toolbox $1000, SimPowerSystems $3000, and SimElectronics $2000.

The MathWorks Maintenance Service sells maintenance contracts for MATLAB and their add-on programs. These contracts provide software updates, bug fixes, and technical support after the first year. For the first year after purchase the maintenance contracts are included. After the first year, the contracts' annual costs are between 10% and 30% the cost of their software.

SYSTEM REQUIREMENTS FOR THE INDIVIDUAL LICENSE PROFESSIONAL VERSION

WINDOWS OPERATING SYSTEM
   Hardware Requirements
      Intel Pentium 4, Intel Celeron, Intel Xeon, Intel Core, or AMD Athlon 64. (Intel Atom, AMD Opteron, or AMD Sempron will run only 32-bit MATLAB software.)
      512 MB of RAM (At least 1024 MB recommended).
      680 MB of hard disk space (without any MATLAB add-on programs).
      16-, 24-, or 32-bit OpenGL capable graphics adapter.
   Software Requirements
      Windows XP Service Pack 2 or Service Pack 3, Windows Server 2003 Service Pack 2 or R2, Windows Vista Service Pack 1 or 2, or Windows 7.
      Microsoft .Word 2002, 2003, or 2007 to run MATLAB Notebook.
      Microsoft Excel 2002, 2003, or 2007 to run Spreadsheet Link EX.

MAC OPERATING SYSTEM
   Hardware Requirements
      Intel Core 2 Duo, Intel Xeon, or Other Intel 64-bit processors. (All other Intel-based Macs will run only 32-bit MATLAB software.)
      1024 MB (At least 2048 MB recommended) (512 MB of RAM [At least 1024 MB recommended] will run 32-bit MATLAB software).
      360 MB of hard disk space (without any MATLAB add-on programs).
   Software Requirements
      Mac OS X 10.5.5 and above or Mac OS X 10.6 and above.

LINUX OPERATING SYSTEM
Hardware Requirements
Intel Pentium 4, Intel Celeron, Intel Xeon, Intel Core, or AMD Athlon 64. (Intel Atom, AMD Opteron, or AMD Sempron will run only 32-bit MATLAB software.)
1024 MB (At least 2048 MB recommended) (512 MB of RAM [At least 1024 MB recommended] will run only 32-bit MATLAB software).
500 MB of hard disk space (without any MATLAB add-on programs).
16-, 24-, or 32-bit OpenGL capable graphics adapter.
Software Requirements
Red Hat Enterprise Linux v.4 and above, OpenSuSE 9.3 and above, Debian 4.0 and above, or Ubuntu 8 and above.

SOLARIS OPERATING SYSTEM (MATLAB release 2009b is the last to support the Solaris SPARC platform)
Hardware Requirements
ultraSPARC
1024 MB (At least 2048 MB recommended).
700 MB of hard disk space (without any MATLAB add-on programs).
24-bit graphics display for Sun Solaris.
Software Requirements
Solaris 10

2.1.4 MULTIPLE USER VERSIONS

MATLAB can also be purchased in versions suitable for multiple users. It is sold with Group Licenses (one administrator overseeing a group of individual licenses), Network Named User Licenses (two or more users going through a server), and Concurrent Licenses (similar to Network Named User License, except that the users do not need to be named). Contact your MATLAB sales representative for details on these.

2.2 ARE THERE MATLAB ADD-ON PROGRAMS RESIDING IN YOUR COMPUTER?

MATLAB has 47 optional add-on programs. One of its add-on programs, Simulink, has 34 optional add-on programs itself. Many studying MATLAB are using copies that they did not purchase themselves. They gained access to it through their employer's copy. Often the employer has purchased add-on programs in addition to the basic MATLAB program. The learner may wonder what add-on options are on his computer.

To check which add-on options are available on your computer: 1) Start MATLAB. 2) Left-click "Help" in the main menu. 3) Left-click "Product Help". 4) The available add-on programs will appear under the "Contents" tab.

2.3 MATLAB TRAINING AND TUTORIALS

MathWorks offers online and classroom training on MATLAB, Simulink, and other products.

Free worthwhile MATLAB training can be found on The MathWorks website at http://www.mathworks.com/academia/student_center/tutorials/launchpad.html:
1) There are 15 free "Getting Started Videos". Video titles of greatest interest to those doing numeric computations are "Getting Started with MATLAB", "Writing a MATLAB program", and "Working with Arrays".
2) There are "Interactive MATLAB Tutorial" videos.
3) There are links to free university-authored MATLAB tutorials.

MathWorks offers 18 classroom courses on MATLAB topics. The classes are usually one or two days long and cost about $600 per day. Classes are taught at different locations around the country and can be taught on-site. The class that would probably be of most interest to beginners is "Fundamentals of MATLAB", a two day course. There are also "Fundamentals…" classes that are made especially for automotive, financial, and aerospace applications.

2.4 HOW POPULAR IS MATLAB?

The MathWorks website states:
1) "Over 1,000,000 engineers and scientists in more than 100 countries, on all seven continents, use MATLAB and Simulink."
2) MATLAB "…products have become fundamental tools for work at the world's most innovative technology companies, government research labs, financial institutions, and at more than 5,000 universities."

MATLAB is a popular program. At the time of writing, Monster.com listed 573 job advertisements that mentioned MATLAB. Furthermore, it is likely that many companies are presuming that new engineering graduates have a background in MATLAB or a similar program.

2.5 MATLAB CLONES

An often mentioned problem with MATLAB is its high price. There are free MATLAB clones that have many of the MATLAB features. Two of the popular clones are GNU Octave and Scilab. Information on GNU Octave can be found on http://www.gnu.org/software/octave/. Information on Scilab can be found on http://www.scilab.org/support/documentation.

2.6 HOW DOES MATLAB COMPARE WITH SIMILAR PROGRAMS?

The software packages Mathcad, Mathematica, and Maple are often compared to MATLAB.

Mathcad was the subject of my earlier book, *Mathcad for Electrical Engineers and Technologists*. Mathcad is owned by PTC (Parametric Technology Corporation). It is often considered to be more "user-friendly" and of an "intuitive and sophisticated calculator" than MATLAB. However, Mathcad is usually considered inferior to MATLAB for sophisticated programming. The cost of Mathcad is much less than that of MATLAB.

Mathematica is a program by Wolfram Research. Its founder stated that "Mathematica is a system for doing mathematics by computer." It was written with the mathematician and scientist in mind. It is good at symbolic calculations. MATLAB only does this with an add-on program. Mathematica is also capable of doing electrical calculations and programming. It is often considered to be between Mathcad and MATLAB in sophisticated programming and graphics capabilities. The cost of Mathematica is also between that of Mathcad and MATLAB.

Maple is a program by Maplesoft a division of Waterloo Maple, Inc. According to its website, "The Maple programming language is designed for solving mathematical problems..." It is similar to Mathematica in capabilities and price.

3.0 EXAMPLE PROBLEMS

The order of the example problems goes roughly from the simplest and most basic to the more complicated and probably less used. Every beginner should do the first couple examples. Later examples can be studied as needed.

Most of the examples can be done with the basic MATLAB program without the use of an add-on program. However, one example requires MATLAB's Optimization Toolbox and another requires MATLAB's Simulink.

3.1 INTERACTIVE MODE (USING MATLAB AS A CALCULATOR)

For the beginner, MATLAB's interactive mode is very useful for trying out program statements.

Problem:
A wire's resistivity is 1.588 Ω/1000 ft. What is the resistance of 32 ft. of this wire?

Solution:
1) Start MATLAB. The default blank screen seen in Figure 3-1-1 should appear. If it does not, go from the "MATLAB Desktop" main menu to "Desktop", then "Desktop Layout", and then left-click "Default".

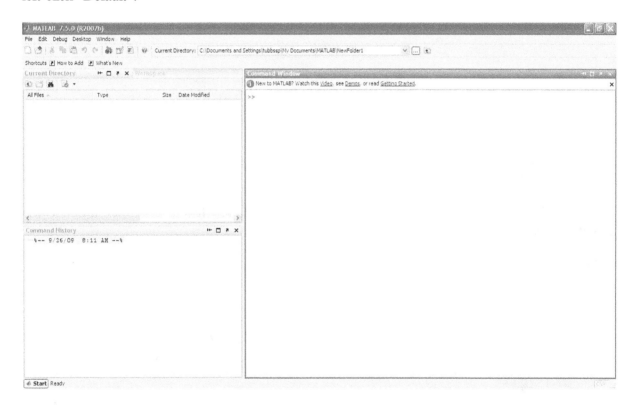

Figure 3-1-1 Blank MATLAB default desktop.

2) In Figure 3-1-1 there are three windows. The "Command Window" is the important one when using MATLAB as a calculator.

3) In the "Command Window" after the ">>" type "R = (1.588/1000)*32" and "Enter".

4) The result is shown in the "Command Window" of Figure 3-1-2.

10

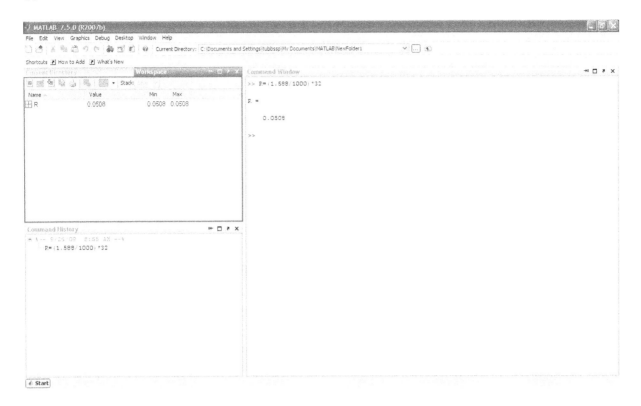

Figure 3-1-2 MATLAB used as a calculator.

5) The Workspace, Current Directory, and Command History windows are not too important here. They will be described in the next example.

3.2 STEADY-STATE ELECTRICAL CIRCUITS

3.2.1 SIMPLE DC CIRCUIT

Problem:

Solve for V1, V2, I, and the power to resistor R2 for the circuit of Figure 3-2-1-1. The following values are given: VS = 10 volts, R1 = 3 Ω, and R2 = 3 Ω.

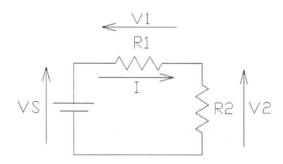

Figure 3-2-1-1 DC supply with a voltage divider.

Solution:

1) Start MATLAB. The default blank screen seen in Figure 3-1-1 should appear. If it does not go to the "MATLAB Desktop" main menu to "Desktop", then "Desktop Layout", and then "Default".

2) In Figure 3-1-1 there are three windows. The "Command Window" is used to enter programs and receive outputs. The "Current Directory" window lists the current directory files. The "Command History" window records all commands typed into the "Command Window".

3) The "Current Directory" window can be toggled to show the "Workspace" window. The "Workspace" window lists the variables used and their values. The "Workspace" window can be seen in Figure 3-2-1-2.

4) The "Editor" window can be selected by left-clicking on the "New M-File" icon beneath "File" in the main menu. This brings up the window shown Figure 3-2-1-2. Programs called "M-Files" are typed into the "Editor" window and later blocked, copied, and pasted to the "Command Window" for running. Programs do not have to be composed with the MATLAB "Editor". They could be created on any text editor and then copied into the "Command Window".

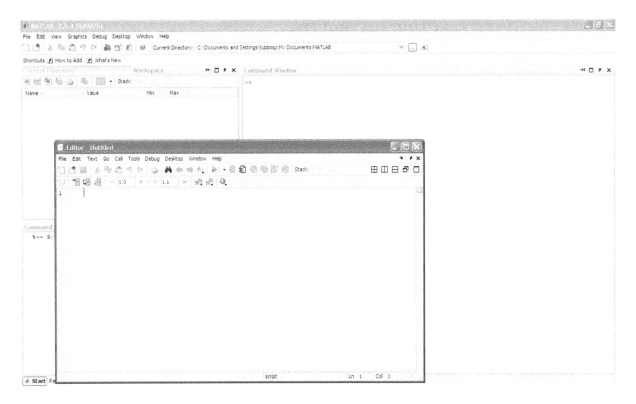

Figure 3-2-1-2 Blank MATLAB desktop showing the "Workspace" window and the "Editor" window.

5) Type the program shown in Figure 3-2-1-3 into the "Editor". Comment lines are preceded by "%"s. MATLAB automatically makes them green. On the first line of the program put the program's file name and a descriptive title. MATLAB does not require these, but it helps keep programs organized.

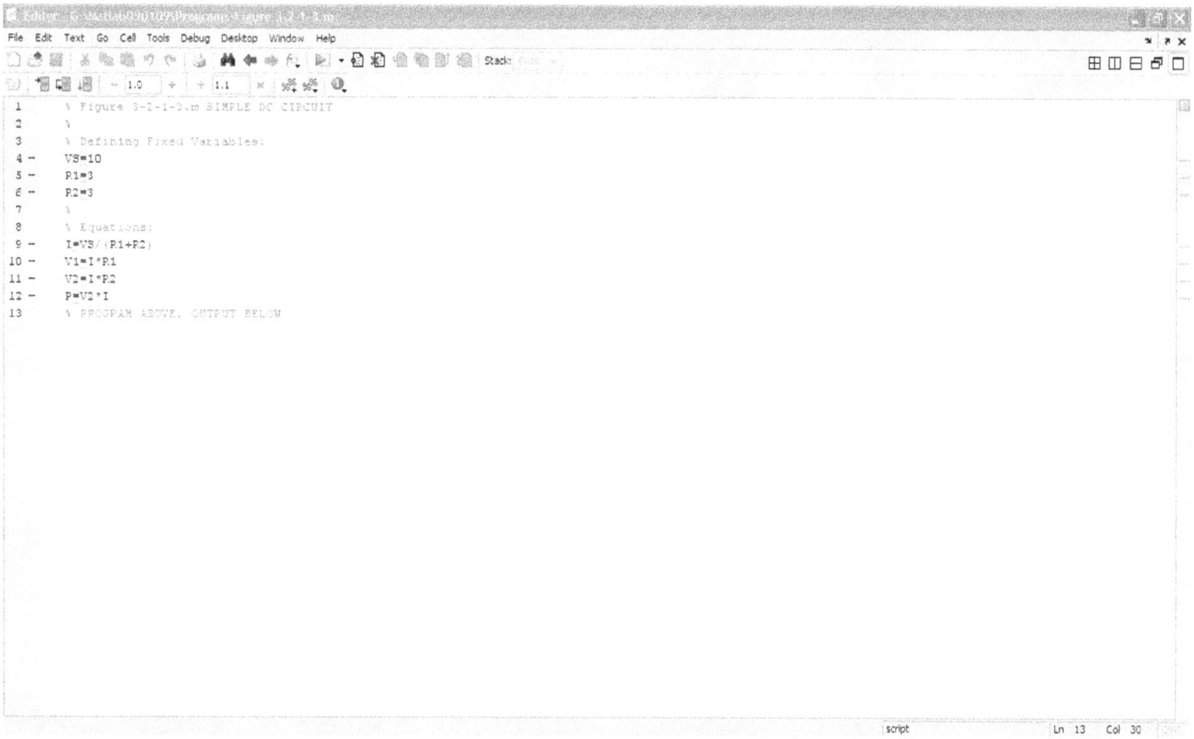

Figure 3-2-1-3 SIMPLE DC CIRCUIT Program typed into the "Editor" window.

6) Save the "Editor" window program. MATLAB automatically puts a ".m" extender after the "Figure 3-2-1-3" file name. As with all computer programs, save often.

7) Block and "Ctrl c" copy the "Editor" window's program.

8) Open the "Command Window" and "Ctrl v" paste the program into it. The "MATLAB Desktop" with the program pasted is shown in Figure 3-2-1-4.

14

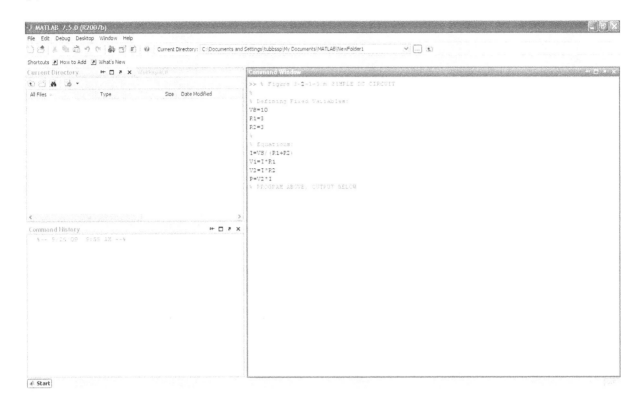

Figure 3-2-1-4 "MATLAB Desktop" with the program pasted into the "Command Window" before it is run.

9) Press "Enter" at the end of the pasted in program and MATLAB will execute the program. The resulting "MATLAB Desktop" is shown in Figure 3-2-1-5. The "Command Window" shows the end of the executed program. The upper parts of the program are still in the "Command Window" and can be accessed by scrolling up the window.

Figure 3-2-1-5 "MATLAB Desktop" after the program has run and the "Workspace" has been selected.

10) Notice how the "Command History" in Figure 3-2-1-5 has recorded all the variables and comments that were initially loaded into the "Command Window". Also, the "Workspace" contains a list of all variables and their values.

11) The "Command Window" contains both the program and the output. The program is single spaced and begins at the top of the window. The output follows and has either one or two spaces between its lines. To make the program and output more distinct a comment line, "% PROGRAM ABOVE, OUTPUT BELOW", has been put at the end of the program. The "Command Window" text is displayed in Figure 3-2-1-6. A grey background has been placed behind it to show that it is different from the book's text.

16

```
% Figure 3-2-1-3.m SIMPLE DC CIRCUIT
%
% Defining Fixed Variables:
VS=10
R1=3
R2=3
%
% Equations:
I=VS/(R1+R2)
V1=I*R1
V2=I*R2
P=V2*I
% PROGRAM ABOVE, OUTPUT BELOW

VS =

    10


R1 =

    3


R2 =

    3


I =

    1.6667


V1 =

    5
```

Continued

V2 =

  5

P =

  8.3333

>>

Figure 3-2-1-6 Text of the "Command Window" after the program has run.

12) If the ".m" file was saved in the "Current Directory", another way of running the program is to type the program's name after the ">>" prompt in the "Command Window". Type only the name not the ".m" extender.

13) The variable values of the "Workspace" are saved by going to the main menu "File" then "Save Workspace As...". Name it "Workspace Figure 3-2-1-4", choose the directory, and save it. MATLAB will automatically put a .mat extender after the file name. This will save the material in the "Workspace" window, but will not save the materials in the windows of the "Current Directory", "Command History", or "Command Window".

14) "Workspace" variables and values of older program runs are retained, unless they are overwritten by new values. Sometimes the retained older values can be a problem. For example, when plotting, MATLAB may use a combination of old program run and new program run array values. This may produce incorrect plots. See Section 3.2.9. For this reason, it is a good idea to clear the "Workspace" before running new programs.

15) Clear the "Command Window", "Command History", and "Workspace" windows by going to the "Edit" menu and selecting "Clear Command Window", "Clear Command History", and "Clear Workspace".

3.2.2 DC MESH CIRCUIT

Problem:

Solve for the currents I1, I2, and I3 in the circuit of Figure 3-2-2-1. The following values are given: VS = 10 volts, R1 = 1 Ω, R2 = 2 Ω, R3 = 3 Ω, R4 = 4 Ω, R5 = 5 Ω, and R6 = 6 Ω.

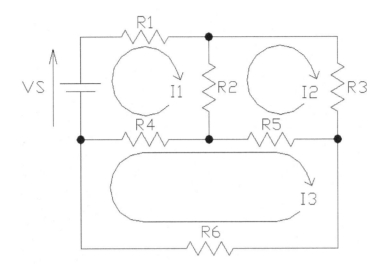

Figure 3-2-2-1 DC mesh circuit.

The simultaneous equations that describe Figure 3-2-2-1 are:

$$VS = I1 \cdot (R1 + R2 + R4) + I2 \cdot (-R2) + I3 \cdot (-R4)$$
$$0 = I1 \cdot (-R2) + I2 \cdot (R2 + R3 + R5) + I3 \cdot (-R5)$$
$$0 = I1 \cdot (-R4) + I2 \cdot (-R5) + I3 \cdot (R4 + R5 + R6)$$

In matrix form the equations would be written:

$$
\begin{pmatrix} VS \\ 0 \\ 0 \end{pmatrix} =
\begin{pmatrix}
(R1+R2+R4) & -R2 & -R4 \\
-R2 & (R2+R3+R5) & -R5 \\
-R4 & -R5 & (R4+R5+R6)
\end{pmatrix}
\bullet
\begin{pmatrix} I1 \\ I2 \\ I3 \end{pmatrix}
$$

Determinants and Cramer's rule can be used to solve for the currents. The determinant equations for $I1$ would be:

$$I1 = \frac{\begin{vmatrix} VS & -R2 & -R4 \\ 0 & (R2+R3+R5) & -R5 \\ 0 & -R5 & (R4+R5+R6) \end{vmatrix}}{\begin{vmatrix} (R1+R2+R4) & -R2 & -R4 \\ -R2 & (R2+R3+R5) & -R5 \\ -R4 & -R5 & (R4+R5+R6) \end{vmatrix}}$$

The currents can also be found from the current matrix. The current matrix is determined by multiplying the inverse of the resistance matrix times the voltage matrix:

$$\begin{pmatrix} I1 \\ I2 \\ I3 \end{pmatrix} = \begin{pmatrix} (R1+R2+R4) & -R2 & -R4 \\ -R2 & (R2+R3+R5) & -R5 \\ -R4 & -R5 & (R4+R5+R6) \end{pmatrix}^{-1} \bullet \begin{pmatrix} VS \\ 0 \\ 0 \end{pmatrix}$$

Here MATLAB solves the equations for currents using three different methods. First, determinants and Cramer's rule are used to solve the equations. Second, MATLAB's "left division" method is used to solve the matrix form of the equations. Third, the equation's resistance matrix is inverted and multiplied by its voltage matrix.

The second, "left division" method and third, "matrix inversion" method, are the most practical for these systems of linear equations. The first method, determinants and Cramer's rule is here just to demonstrate MATLAB's determinant handling capabilities.

20

3.2.2.1 DC Mesh Circuit Solution using Determinants

MATLAB's determinant solving ability is demonstrated in this example.

1) MATLAB uses a similar format to that used in Section 3.2.1. However, here determinants are used with Cramer's rule. The program is in Figure 3-2-2-1-1.

```
% Figure 3-2-2-1-1.m DC MESH ANALYSIS WITH DETERMINANTS
%
%Defining fixed variables:
VS=10
R1=1
R2=2
R3=3
R4=4
R5=5
R6=6
% Equations
D=det([R1+R2+R4 -R2 -R4; -R2 R2+R3+R5 -R5; -R4 -R5 R4+R5+R6])
I1N=det([VS -R2 -R4; 0 R2+R3+R5 -R5; 0 -R5 R4+R5+R6])
I2N=det([R1+R2+R4 VS -R4; -R2 0 -R5; -R4 0 R4+R5+R6])
I3N=det([R1+R2+R4 -R2 VS; -R2 R2+R3+R5 0; -R4 -R5 0])
I1=I1N/D
I2=I2N/D
I3=I3N/D
V1check=I1*(R1+R2+R4)+I2*(-R2)+I3*(-R4)
V2check=I1*(-R2)+I2*(R2+R3+R5)+I3*(-R5)
V3check=I1*(-R4)+I2*(-R5)+I3*(R4+R5+R6)
% PROGRAM ABOVE, OUTPUT BELOW

VS =

 10


R1 =

 1
```

Continued

R2 =

2

R3 =

3

R4 =

4

R5 =

5

R6 =

6

D =

575

I1N =

1250

I2N =

500

Continued

I3N =

  500

I1 =

  2.1739

I2 =

  0.8696

I3 =

  0.8696

V1check =

  10.0000

V2check =

  0

V3check =

  0

Figure 3-2-2-1-1 MATLAB DC MESH ANALYSIS using determinants with input data, output data, determinant equations, and check equations.

    2) Note how the determinant, D, is written in MATLAB. The matrix for the system of equations is written inside square brackets on a single line with elements separated by spaces and with rows separated by semicolons, [R1+R2+R4 -R2 -R4; -R2 R2+R3+R5 -R5; -R4 -R5 R4+R5+R6]. Then the MATLAB determinant function "det(…)" is written around the matrix.

3.2.2.2 DC Mesh Circuit Solution using MATLAB's Matrix Left Division

MATLAB's ability to solve matrices with its matrix "left division" feature is demonstrated in this example.

1) This is similar to the solution in 3.2.2.1 except that MATLAB's "left division" feature is used. Here the resistance matrix is "left divided" by the one column voltage matrix. Notice how MATLAB uses single variable names, R and v, to represent entire matrices. The MATLAB symbol for "left division" is "\". The result is the values of the currents in a one column current matrix. The program is in Figure 3-2-2-2-1.

```
% Figure 3-2-2-2-1.m DC MESH ANALYSIS WITH LEFT DIVISION
%
%Defining fixed variables:
VS=10
R1=1
R2=2
R3=3
R4=4
R5=5
R6=6
% Equations
R=[R1+R2+R4 -R2 -R4; -R2 R2+R3+R5 -R5; -R4 -R5 R4+R5+R6]
v=[VS; 0; 0]
I=R\v
V1check=I(1)*(R1+R2+R4)+I(2)*(-R2)+I(3)*(-R4)
V2check=I(1)*(-R2)+I(2)*(R2+R3+R5)+I(3)*(-R5)
V3check=I(1)*(-R4)+I(2)*(-R5)+I(3)*(R4+R5+R6)
% PROGRAM ABOVE, OUTPUT BELOW

VS =

   10


R1 =

   1
```

Continued

24

R2 =

  2

R3 =

  3

R4 =

  4

R5 =

  5

R6 =

  6

R =

$$\begin{array}{rrr} 7 & -2 & -4 \\ -2 & 10 & -5 \\ -4 & -5 & 15 \end{array}$$

v =

$$\begin{array}{r} 10 \\ 0 \\ 0 \end{array}$$

Continued

I =

2.1739
0.8696
0.8696

V1check =

10.0000

V2check =

0

V3check =

0

Figure 3-2-2-2-1 MATLAB DC MESH ANALYSIS using "left division" with input data, output data, and check equations.

2) Notice that the resultant currents are in a one column "I" matrix. When they are checked they are accessed by I's followed by a number in parentheses. That number is the order the values would have if the matrix was laid out flat. It is also possible to represent the currents by two number coordinates. This is done in Section 3.2.2.3.

26

3.2.2.3 DC Mesh Circuit Solution using MATLAB's Matrix Inversion

MATLAB's ability to solve matrices using "matrix inversion" is demonstrated in this example.

1) This is very similar to the solution in 3.2.2.2 except that MATLAB's "matrix inversion" feature, "inv(…)" and matrix multiplication features are used. Here the resistance matrix is inverted and then multiplied by the one column voltage matrix. The results are the values of the currents in a one column current matrix. The program is in Figure 3-2-2-3-1.

```
% Figure 3-2-2-3-1.m DC MESH ANALYSIS WITH MATRIX INVERSION
%
%Defining fixed variables:
VS=10
R1=1
R2=2
R3=3
R4=4
R5=5
R6=6
% Equations
R=[R1+R2+R4 -R2 -R4; -R2 R2+R3+R5 -R5; -R4 -R5 R4+R5+R6]
v=[VS; 0; 0]
RI=inv(R)
I=RI*v
V1check=I(1,1)*(R1+R2+R4)+I(2,1)*(-R2)+I(3,1)*(-R4)
V2check=I(1,1)*(-R2)+I(2,1)*(R2+R3+R5)+I(3,1)*(-R5)
V3check=I(1,1)*(-R4)+I(2,1)*(-R5)+I(3,1)*(R4+R5+R6)
% PROGRAM ABOVE, OUTPUT BELOW

VS =

   10


R1 =

    1
```

Continued

R2 =

  2

R3 =

  3

R4 =

  4

R5 =

  5

R6 =

  6

R =

$$\begin{array}{rrr} 7 & -2 & -4 \\ -2 & 10 & -5 \\ -4 & -5 & 15 \end{array}$$

v =

$$\begin{array}{r} 10 \\ 0 \\ 0 \end{array}$$

Continued

RI =

```
0.2174   0.0870   0.0870
0.0870   0.1548   0.0748
0.0870   0.0748   0.1148
```

I =

```
2.1739
0.8696
0.8696
```

V1check =

```
10
```

V2check =

```
-1.7764e-015
```

V3check =

```
-3.5527e-015
```

Figure 3-2-2-3-1 MATLAB DC MESH ANALYSIS using "matrix inversion" and multiplication with input data, output data, and check equations.

2) In the check equations the currents are accessed by I's followed by two numbers in parentheses, the first number for the row and the second for the column. It is also possible to represent the currents by single numbers in parentheses. Single number representation was done in Section 3.2.2.2.

3.2.3 SIMPLE AC PHASOR CIRCUIT

MATLAB represents a complex number with a single variable name. To get the real and imaginary values of a complex number, the "real(…)" and "imag(…)" functions are used. To get the magnitude and angle of a complex number "abs(…)" and "angle(…)" functions are used. Angles are in radians.

MATLAB represents input imaginary numbers with "i" and "j" designations. By default, it designates output imaginary numbers with i's.

Problem:

Determine the phasor values **V1** and **I** and the power to resistor R1 for the circuit of Figure 3-2-3-1. The following values are given: VS = 10 volts rms, XC1 = 3 Ω, and R1 = 3 Ω.

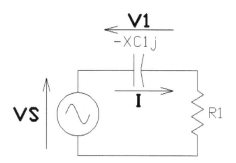

Figure 3-2-3-1 AC supply with a RC voltage divider.

MATLAB's use of "complex number", "abs", and "angle" features are demonstrated in this example.

Solution:

1) In this book's text and figures, phasor values will be indicated by larger bold print. For example, the phasor value for VS is **VS**. In the MATLAB program it is not possible to change the font of individual characters.

2) Write the program of Figure 3-2-3-2.

```
% Figure 3-2-3-2.m SIMPLE AC PHASOR CIRCUIT ANALYSIS
% Defining Fixed Variables
VS=10
R1=3
XC1=3
% Equations
I=VS/(R1+XC1*(-j))
V1=XC1*I*(-j)
magI=abs(I)
angleradI=angle(I)
angledegI=57.2958*angleradI
magV1=abs(V1)
angleradV1=angle(V1)
angledegV1=57.2958*angleradV1
P=(magI^2)*R1
% PROGRAM ABOVE, OUTPUT BELOW

VS =

    10


R1 =

    3


XC1 =

    3


I =

   1.6667 + 1.6667i


V1 =

   5.0000 - 5.0000i
```

Continued

magI =

  2.3570

angleradI =

  0.7854

angledegI =

  45.0000

magV1 =

  7.0711

angleradV1 =

  -0.7854

angledegV1 =

  -45.0000

P =

  16.6667

Figure 3-2-3-2 MATLAB SIMPLE AC PHASOR CIRCUIT ANALYSIS using complex numbers with input data, output data, and check equations.

3.2.4 AC PHASOR MESH CIRCUIT

Problem:

Solve for the phasor currents **I1**, **I2**, and **I3** in the circuit of Figure 3-2-4-1. The following values are given: VS = 10 volts rms, R1 = 1 Ω, R2 = 4 Ω, XL1 = 2 Ω, XL2= 5 Ω, XC1 = 3 Ω, and XC2 = 6 Ω.

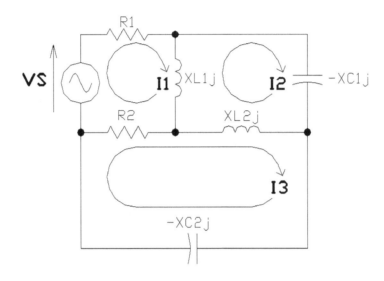

Figure 3-2-4-1 AC mesh circuit.

Solution:

1) The equations for this mesh circuit are:

$$\mathbf{VS} = \mathbf{I1}{\cdot}(R1 + R2 + XL1j) + \mathbf{I2}{\cdot}(\text{-}XL1j) + \mathbf{I3}{\cdot}(\text{-}R2)$$
$$0 = \mathbf{I1}{\cdot}(\text{-}XL1j) + \mathbf{I2}{\cdot}[(XL1 + XL2 - XC1)j] + \mathbf{I3}{\cdot}(\text{-}XL2j)$$
$$0 = \mathbf{I1}{\cdot}(\text{-}R2) + \mathbf{I2}{\cdot}(\text{-}XL2j) + \mathbf{I3}{\cdot}[R2 + (XL2 - XC2)j]$$

2) This circuit can be solved by the same methods as were used in the analysis of the DC mesh circuit, determinants, "left division", and multiplication of the inverse impedance matrix by the voltage matrix. The "left division" method of solution is used in Figure 3-2-4-2.

```
% Figure 3-2-4-2.m SIMPLE AC PHASOR CIRCUIT ANALYSIS
% Defining Fixed Variables
VS=10
R1=1
R2=4
XC1=3
XC2=6
XL1=2
XL2=5
% Equations
Z=[R1+R2+XL1*(j) XL1*(-j) -R2; XL1*(-j) (XL1+XL2-XC1)*j XL2*(-j);
   -R2 XL2*(-j) R2+(XL2-XC2)*(j)]
v=[VS; 0; 0]
I=Z\v
I1=I(1)
MagI1=abs(I1)
AngleI1=57.2958*angle(I1)
I2=I(2)
MagI2=abs(I2)
AngleI2=57.2958*angle(I2)
I3=I(3)
MagI3=abs(I3)
AngleI3=57.2958*angle(I3)
V1Check=I(1)*(R1+R2+XL1*j)+I(2)*(XL1*(-j))+I(3)*(-R2)
V2Check=I(1)*(XL1*(-j))+I(2)*((XL1+XL2-XC1)*j)+I(3)*(XL2*(-j))
V3Check=I(1)*(-R2)+I2*(XL2*(-j))+I3*(R2+(XL2-XC2)*j)
% PROGRAM ABOVE, OUTPUT BELOW

VS =

    10


R1 =

    1


R2 =

    4
```

Continued

34

XC1 =

　3

XC2 =

　6

XL1 =

　2

XL2 =

　5

Z =

$$
\begin{array}{lll}
5.0000 + 2.0000i & 0 - 2.0000i & -4.0000 \\
0 - 2.0000i & 0 + 4.0000i & 0 - 5.0000i \\
-4.0000 & 0 - 5.0000i & 4.0000 - 1.0000i
\end{array}
$$

v =

$$
\begin{array}{l}
10 \\
0 \\
0
\end{array}
$$

I =

$$
\begin{array}{l}
1.4781 + 0.2705i \\
0.4895 + 1.1758i \\
-0.1997 + 0.8324i
\end{array}
$$

Continued

I1 =

  1.4781 + 0.2705i

MagI1 =

  1.5027

AngleI1 =

  10.3694

I2 =

  0.4895 + 1.1758i

MagI2 =

  1.2736

AngleI2 =

  67.3973

I3 =

  -0.1997 + 0.8324i

MagI3 =

  0.8560

Continued

AngleI3 =

103.4883

V1Check =

10.0000 + 0.0000i

V2Check =

2.6645e-015 -6.6613e-016i

V3Check =

-1.1102e-016 +4.4409e-016i

Figure 3-2-4-2 MATLAB AC PHASOR MESH ANALYSIS CIRCUIT with input data, equations, and solutions.

3.2.5 AC INDUCTION MOTOR ANALYSIS

Problem:

An AC three-phase 5 hp induction motor powers a fan. The power output of the motor is assumed to be a constant 5 hp. Using MATLAB and the motor's equivalent circuit determine the motor's current, power loss, and slip at input voltages of 480 and 375 Vrms.

The values for the equivalent circuit of the motor are R1 = 1.6 Ω, R2 = 1.4 Ω, RM= 980 Ω, XM = 120 Ω, X = 6.5 Ω. **VS** is assumed to be at angle 0. These can be seen in Figure 3-2-5-1.

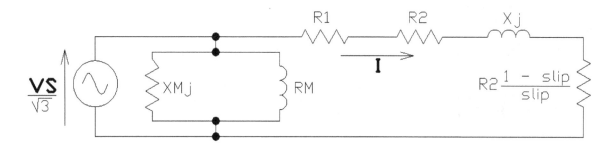

Figure 3-2-5-1 AC induction motor phasor equivalent circuit of one line-to-neutral phase.

The system of equations that describes this problem is non-linear.

(i)   Motor output power (watts) $PO = 3 \cdot |\mathbf{I}|^2 \cdot R2 \cdot (1\text{-slip})/\text{slip}$

(ii)   Rotor equivalent circuit current (amps) $\mathbf{I} = \{\mathbf{VS}/[\text{SQRT}(3)]\}/(R1 + R2/\text{slip} + Xj)$

(iii)   Motor power loss (watts) $PL = 3 \cdot [|\mathbf{I}|^2 \cdot (R1 + R2) + |\mathbf{VS}/[\text{SQRT}(3)]|^2/RM]$

(iv)   Total motor current (amps) $\mathbf{IT} = \mathbf{I} + \{\mathbf{VS}/[\text{SQRT}(3)]\}/RM + \{\mathbf{VS}/[\text{SQRT}(3)]\}/(XMj)\}$

MATLAB can solve the equations itself with a try and test conditional program or with the aid of the MATLAB Optimization Toolbox.

3.2.5.1 Solution using a Try and Test Program

  With the try and test method, potential solution values are put into the equations and the equations checked to see how well they are solved. The equations being solved are equated to zero. Potential solution values are substituted into the equated equations. The equated equations are checked to see how different they are from zero. Solution values that produce the smallest differences are the closest to correct.

  MATLAB's ability to evaluate equations over a range of data, to conditionally check and store data, and to suppress display of output with a ";" are demonstrated in this example.

Solution:
  1) The equations (i) and (ii) are combined to eliminate the **I** variable and the result equated to zero..

$$0 = PO - 3 \cdot \left| [\mathbf{VS}/(SQRT(3)]/(R1 + R2/slip + Xj) \right|^2 \cdot R2 \cdot (1 - slip)/slip$$

  2) A program is written that solves for PO - 3 · $\left| [\mathbf{VS}/(SQRT(3)]/(R1 + R2/slip + Xj) \right|^2$ ·R2·(1 - slip)/slip for each slip value. The value found is called "diff".

  3) The program compares each newly found "diff" to each previously found "diff". If the newly found absolute value of "diff" is less than its previous absolute value, its value and its corresponding slip value are stored.

  4) After all slip values in the designated range have been tried, the slip value that produces the smallest "diff" is used to calculate **I**, PL, and **IT**.

  5) The program is shown in Figure 3-2-5-1-1.

```
% Figure 3-2-5-1-1.m AC INDUCTION MOTOR WITH TRY AND TEST WITH 480V
sliptry=.0001
diffmin= abs(5*746 - (3*(abs((480/(3^.5))/(1.6 + 1.4/sliptry + 6.5j)))^2)*1.4*(1-sliptry)/sliptry)
for sliptry=.001:.001:.5
diff = abs(5*746 - (3*(abs((480/(3^.5))/(1.6 + 1.4/sliptry + 6.5j)))^2)*1.4*(1-sliptry)/sliptry)
if diff<diffmin
   diffmin=diff
   slip=sliptry
end
end
slip=slip
I=(480/(3^.5))/(1.6+1.4/slip+6.5j)
PL=3*((abs(I)^2)*(1.6+1.4)+375^2/(3*980))
IT=I+375/(3^.5*980)+375/(3^.5*120j)
% The solution check follows
PowerHpCheck=3*abs(I)^2*1.4*(1-slip)/(slip*746)
MotorVoltageCheck=3^.5*I*(1.6+1.4/slip+6.5j)
% PROGRAM ABOVE, OUTPUT BELOW


sliptry =

  1.0000e-004


diffmin =

  3.7135e+003


diff =

  3.5660e+003


diffmin =

  3.5660e+003
```

Continued

40

slip =

  1.0000e-003

<Note: Here 8 pages of output slip, diff, and diffmin values were not printed to save paper>

diff =

  13.9889

diffmin =

  13.9889

slip =

  0.0250

<Note: Here 54 pages of diff values were not printed to save paper>

slip =

  0.0250

I =

  4.7508 - 0.5361i

PL =

  349.2086

Continued

IT =

  4.9717 - 2.3403i

PowerHpCheck =

  5.0188

MotorVoltageCheck =

  480.0000

Figure 3-2-5-1-1 AC INDUCTION MOTOR WITH TRY AND TEST WITH 480 V.

6) Notice that the range of slips was only from .001 to .5 rather than .001 to 1.0. At the unreasonably large slip of .6 the equations equate closer to zero than they do at the reasonable slip of .025. The large slip solution was avoided by simply not evaluating in that range. The MATLAB programmer should be alert for unreasonable answers.

7) When a MATLAB program is run it displays every variable that is followed by an equals sign. In the program of Figure 3-2-5-1-1, 62 pages of unneeded data were displayed by the program that were not copied into this book. A simple way of eliminating unneeded data is to put a ";" to the right of the its equation. For example, "diffmin=diff" would become "diffmin=diff;". The program of Figure 3-2-5-1-1 is rewritten and run with ";"s on all equations except for the output equations.

```
% Figure 3-2-5-1-1.m AC INDUCTION MOTOR WITH TRY AND TEST WITH 480V
sliptry=.0001;
diffmin= abs(5*746 - (3*(abs((480/(3^.5))/(1.6 + 1.4/sliptry + 6.5j)))^2)*1.4*(1-sliptry)/sliptry);
for sliptry=.001:.001:.5
diff= abs(5*746 - (3*(abs((480/(3^.5))/(1.6 + 1.4/sliptry + 6.5j)))^2)*1.4*(1-sliptry)/sliptry);
if diff<diffmin
    diffmin=diff;
    slip=sliptry;
end
end
slip=slip
I=(480/(3^.5))/(1.6+1.4/slip+6.5j)
PL=3*((abs(I)^2)*(1.6+1.4)+375^2/(3*980))
IT=I+375/(3^.5*980)+375/(3^.5*120j)
```

Continued

42

% The solution check follows
PowerHpCheck=3*abs(I)^2*1.4*(1-slip)/(slip*746)
MotorVoltageCheck=3^.5*I*(1.6+1.4/slip+6.5j)
% PROGRAM ABOVE, OUTPUT BELOW

slip =

  0.0250

I =

  4.7508 - 0.5361i

PL =

 349.2086

IT =

  4.9717 - 2.3403i

PowerHpCheck =

  5.0188

MotorVoltageCheck =

 480.0000

Figure 3-2-5-1-2 AC INDUCTION MOTOR WITH TRY AND TEST WITH 480 V and ";"s put after all equations except those for output.

8) The program is redone at 375V by substituting 480 by 375 and decreasing the range of the evaluated slips to .001 to .4. The slip range needs to be less than .45. At a slip of .45 (an unreasonably large slip) the equations come closer to equating to zero than at the reasonable slip of .045. The program output is in Figure 3-2-5-1-3.

slip =

 0.0450

I =

 6.3673 - 1.2652i

PL =

 522.7879

IT =

 6.5882 - 3.0695i

PowerHpCheck =

 5.0354

MotorVoltageCheck =

 375.0000

Figure 3-2-5-1-3 AC INDUCTION MOTOR WITH TRY AND TEST WITH 375 V program output.

3.2.5.2 Solution using the Optimization Toolbox "fsolve" Subroutine

The system can be solved by using MATLAB with its add-on Optimization Toolbox "fsolve" subroutine.

MATLAB's Optimization Toolbox "fsolve" uses numerical analysis methods to iteratively determine the best solution for systems of equations. "fsolve" finds the values of a function that will make the function equate to zero. For example, the equation $4 = x^2$ would be converted to $0 = 4 - x^2$ to apply "fsolve" to it. "fsolve" is of greatest use with non-linear equations.

Two programs are necessary to use "fsolve", a main program and a subroutine. The main program calls the subroutine. It is necessary to store the subroutine in the MATLAB default directory, so that the main program can find it.

There are many options with "fsolve". Different numerical techniques, tolerances, and output formats can be selected. In this example, only the number of evaluations is specified, the other options are left at default values.

Solution:

1) Solving the equations (i) and (ii) for $\mathbf{I}$ and slip provides values to solve the last two equations.

2) To make the output of the MATLAB program more compact the equations (i) and (ii) have numerical values for PO, R2, $\mathbf{VS}$, R1, and X inserted into them.

$5 \cdot 746 = 3 \cdot |\mathbf{I}|^2 \cdot 1.4 \cdot (1\text{-slip})/\text{slip}$

$\mathbf{I} = \{480/[\text{SQRT}(3)]\}/(1.6 + 1.4/\text{slip} + 6.5j)$

3) To make the equations ready for use in an Optimization Toolbox "fsolve" subroutine, the equations are equated to zero.

$5 \cdot 746 - 3 \cdot |\mathbf{I}|^2 \cdot 1.4 \cdot (1 - \text{slip})/\text{slip} = 0$

$\mathbf{I} - \{480/[\text{SQRT}(3)]\}/(1.6 + 1.4/\text{slip} + 6.5j) = 0$

4) Enter the "fsolve" subroutine program shown in Figure 3-2-5-2-1. Notice how slip and current, $\mathbf{I}$, are redefined as a(1) and a(2). This is so that the slip and current can be written into a single matrix. This subroutine is stored in the MATLAB default directory by its name "motorfun.m", so that the main program can find it.

```
% motorfun.m AC INDUCTION MOTOR FSOLVE SUBROUTINE WITH 480V
% a(1)=slip
% a(2)=I
function F = motorfun(a)
% Equations
F = [5*746-3*(abs(a(2)))^2*1.4*(1-a(1))/a(1);
a(2)-(480/(3^.5))/(1.6+1.4/a(1)+6.5j)]
```

Figure 3-2-5-2-1 MATLAB AC INDUCTION MOTOR FSOLVE SUBROUTINE WITH 480V.

5) Enter the main program shown in Figure 3-2-5-2-2.

```
% Figure 3-2-5-2-2.m AC INDUCTION MOTOR WITH FSOLVE AND 480V
% a(1) = slip
% a(2) = I (current, amps)
% a0 contains the initial guess values for a(1) and a(2)
a0=[.02; 5-.6j]
%a0=[.025; 4.7376-.57j]
% fsolve calls up the program subroutine, motorfun.m, from
% the MATLAB default directory and solves for a(1)and a(2).
options=optimset('MaxFunEval',5)
a=fsolve(@motorfun,a0,options)
%Other equations that are solved
PL=3*((abs(a(2))^2)*(1.6+1.4)+480^2/(3*980))
IT=a(2)+480/(3^.5*980)+480/(3^.5*120j)
% The solution check follows
PowerHpCheck=3*abs(a(2))^2*1.4*(1-a(1))/(a(1)*746)
MotorVoltageCheck=3^.5*a(2)*(1.6+1.4/a(1)+6.5j)
% PROGRAM ABOVE, OUTPUT BELOW

a0 =

0.0200
5.0000 - 0.6000i
```

Continued

46

options =

```
                Display: []
           MaxFunEvals: 5
                MaxIter: []
                 TolFun: []
                   TolX: []
            FunValCheck: []
              OutputFcn: []
               PlotFcns: []
        ActiveConstrTol: []
         BranchStrategy: []
        DerivativeCheck: []
            Diagnostics: []
          DiffMaxChange: []
          DiffMinChange: []
      GoalsExactAchieve: []
             GradConstr: []
                GradObj: []
                Hessian: []
               HessMult: []
            HessPattern: []
             HessUpdate: []
         InitialHessType: []
       InitialHessMatrix: []
               Jacobian: []
              JacobMult: []
           JacobPattern: []
             LargeScale: []
     LevenbergMarquardt: []
         LineSearchType: []
               MaxNodes: []
             MaxPCGIter: []
             MaxRLPIter: []
             MaxSQPIter: []
                MaxTime: []
          MeritFunction: []
              MinAbsMax: []
    NodeDisplayInterval: []
     NodeSearchStrategy: []
```

Continued

```
          NonlEqnAlgorithm: []
          NoStopIfFlatInfeas: []
      PhaseOneTotalScaling: []
             Preconditioner: []
          PrecondBandWidth: []
             RelLineSrchBnd: []
     RelLineSrchBndDuration: []
           ShowStatusWindow: []
                    Simplex: []
                     TolCon: []
                     TolPCG: []
                   TolRLPFun: []
                 TolXInteger: []
                   TypicalX: []
```

F =

1.0e+003 *

-1.4891
 0.0012 - 0.0003i

F =

1.0e+003 *

-1.4891
 0.0012 - 0.0003i

F =

1.0e+003 *

-1.4891
 0.0012 - 0.0003i

Continued

48

F =

  1.0e+002 *

  -4.2112 + 0.3317i
   0.0002 + 0.0005i

F =

  1.0e+002 *

  -4.2112 + 0.3317i
   0.0002 + 0.0005i

F =

  1.0e+002 *

  -4.2112 + 0.3317i
   0.0002 + 0.0005i

Maximum number of function evaluations reached:
increase options.MaxFunEvals.

a =

  0.0278 + 0.0002i
  5.2850 - 0.5721i

PL =

  489.4306

IT =

  5.5678 - 2.8815i

Continued

PowerHpCheck =

  5.5645 - 0.0445i

MotorVoltageCheck =

  4.8174e+002 +4.4294e+000i

Figure 3-2-5-2-2 MATLAB AC INDUCTION MOTOR ANALYSIS WITH FSOLVE AND 480V main program.

6) Notice the equations for PL, IT, and the check equations were not involved in the numerical methods used in the "fsolve" subroutine.

7) Notice the maximum number of evaluations was set at 5. This was noted in the program and the output. In the output MATLAB also lists other possible options and makes the statement that the "Maximum number of function evaluations reached: increase options. MaxFunEvals." A more accurate answer could have been obtained with more evaluations. However, the check equations indicate that nearly correct values have been found. Furthermore, the found values seem reasonable for a motor. Check equations are good to use on any MATLAB program, but are especially useful when numerical methods like those of the Optimization Toolbox are used.

8) The 480 V values in the subroutine and main program are replaced with 375 V, the maximum number of evaluations option is not used, and the program is rerun. The programs and results are shown in Figures 3-2-5-2-3 and 3-2-5-2-4.

```
% motorfun.m AC INDUCTION MOTOR FSOLVE SUBROUTINE WITH 375V
% a(1)=slip
% a(2)=I
function F = motorfun(a)
% Equations
F = [5*746-3*(abs(a(2)))^2*1.4*(1-a(1))/a(1);
375-(3^.5)*a(2)*(1.6+1.4/a(1)+6.5j)];
```

Figure 3-2-5-2-3 MATLAB AC INDUCTION MOTOR FSOLVE SUBROUTINE WITH 375V.

50

>> % Figure 3-2-5-2-4.m AC INDUCTION MOTOR WITH FSOLVE AND 375V
% a(1) = slip
% a(2) = I (current, amps)
% a0 contains the initial guess values for a(1) and a(2)
a0=[.02; 5-.6j]
% fsolve calls up the program subroutine, motorfun.m, from
% the MATLAB default directory and solves for a(1)and a(2).
a=fsolve(@motorfun,a0)
%Other equations that are solved
PL=3*((abs(a(2))^2)*(1.6+1.4)+375^2/(3*980))
IT=a(2)+375/(3^.5*980)+375/(3^.5*120j)
% The solution check follows
PowerHpCheck=3*abs(a(2))^2*1.4*(1-a(1))/(a(1)*746)
MotorVoltageCheck=3^.5*a(2)*(1.6+1.4/a(1)+6.5j)
% PROGRAM ABOVE, OUTPUT BELOW

a0 =

  0.0200
  5.0000 - 0.6000i

F =

  1.0e+003 *

  -1.4891
   0.0020 - 0.0003i

<Note: 83 "F =…" values were not printed here to save paper>

F =

  0.0000 + 0.0021i
  0.5414 + 1.5867i

Optimizer appears to be converging to a point which is not a root.
 Norm of relative change in X is less than max(options.TolX^2,eps) but
 sum-of-squares of function values is greater than or equal to sqrt(options.TolFun)
 Try again with a new starting guess.

Continued

a =

  0.0350 + 0.0000i
  5.6154 + 0.7950i

PL =

  432.9774

IT =

  5.8363 - 1.0093i

PowerHpCheck =

  5.0000 - 0.0000i

MotorVoltageCheck =

  3.9620e+002 +1.2057e+002i

Figure 3-2-5-2-4 MATLAB AC INDUCTION MOTOR WITH FSOLVE AND 375V.

     9) Here MATLAB states, "Optimizer appears to be converging to a point which is not a root…. Try again with a new starting guess." Also the MotorVoltageCheck produces a voltage value significantly different from the expected 375. Choosing a better starting guess values would improve the output. Also, using some of the other MATLAB Optimization Toolbox options would help find a more accurate answer.

3.2.6 AC INDUCTION MOTOR 2-D PLOT OF EFFICIENCY VERSUS SLIP

Problem:

Use MATLAB to create a plot of efficiency versus slip for the motor of Section 3.2.5. Use an input voltage of 480 V rms and the same equivalent circuit as in Section 3.2.5.

MATLAB's plot feature, range variable feature, and array arithmetic operators are demonstrated in this example.

Solution:

1) The equations needed are:

Rotor equivalent circuit current (amps) $\mathbf{I} = \mathbf{VS}/[\text{SQRT}(3)\cdot(R1 + R2/slip + Xj)]$

Motor output power PO (watts) $= 3\cdot|\mathbf{I}|^2\cdot R2\cdot(1 - slip)/slip$

Motor power loss (watts) $PL = 3\cdot[|\mathbf{VS}|^2/(3\cdot RM) + |\mathbf{I}|^2\cdot(R1 + R2)]$

Efficiency (pu) $EFF = PO/(PO + PL)$

2) Enter the equations into MATLAB as shown in Figure 3-2-6-1.

```
% Figure 3-2-6-1.m INDUCTION MOTOR EFFICIENCY VERSUS SLIP
% Defining Fixed Variables
VS=480
R1=1.6
R2=1.4
RM=980
XM=120
X=6.5
% Define range of slip
slip = .001: .02: .3;
I=VS./((3^.5)*(R1+R2./slip + X*(j)));
PO=3.*(abs(I)).^2*R2.*(1-slip)./slip
PL=3.*((VS^2)/(3*RM)+((abs(I)).^2)*(R1+R2))
EFF=PO./(PO+PL)
plot(slip,EFF)
% PROGRAM ABOVE, OUTPUT BELOW
```

Continued

VS =

  480

R1 =

  1.6000

R2 =

  1.4000

RM =

  980

XM =

  120

X =

  6.5000

PO =

1.0e+004 *

Columns 1 through 9

  0.0164   0.3198   0.5716   0.7698   0.9175   1.0206   1.0863   1.1218   1.1338

Columns 10 through 15

  1.1280   1.1091   1.0807   1.0457   1.0064   0.9645

Continued

54

PL =

1.0e+003 *

Columns 1 through 9

  0.2355   0.3821   0.7587   1.3067   1.9681   2.6922   3.4395   4.1810   4.8975

Columns 10 through 15

  5.5771   6.2137   6.8047   7.3502   7.8519   8.3122

EFF =

Columns 1 through 9

  0.4106   0.8933   0.8828   0.8549   0.8234   0.7913   0.7595   0.7285   0.6984

Columns 10 through 15

  0.6692   0.6409   0.6136   0.5872   0.5617   0.5371

Figure 3-2-6-1 MATLAB program to produce data for a plot of an AC induction motor's "efficiency" versus "slip".

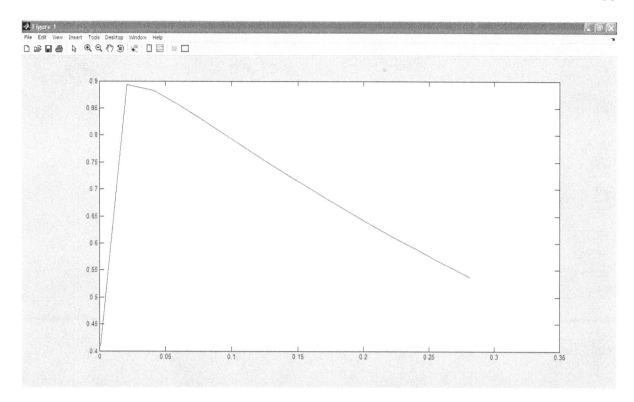

Figure 3-2-6-2 Plot of an AC induction motor's "efficiency" versus "slip".

3) The program generates an array of "slip" data with the "slip = .001: .02: .3;" statement. With this statement MATLAB selects values for "slip"s of .001, 001+.02, .001+.02+.02, to .281. Their values can be accessed as slip(1), slip(2), slip(3),….slip(15).

4) There are special arithmetic operators in these equations. They are meant for use with elements of arrays. For example, look at "I=VS./((3^.5)*(R1+R2./slip + X*(j)));". In this statement the periods to the left of the "/"'s make the calculation consider one slip element (i.e. slip(3)) at a time and then create one I element (i.e. I(3)). Special arithmetic statements are used in the other program lines as well. "./" is array division. ".*" is array multiplication. ".^" is an array exponent. The result of all this is that arrays of values are created for slip, I, PO, PL, and EFF.

5) The "plot(slip,EFF)" statement created a plot of EFF versus slip. Slip is the x axis. EFF is the y axis. The plot appeared over the "Command Window" when the program was run. If desired, the plot can be saved.

6) The axes and plot can be labeled using the menus in the "Figure 1" window. To label the x axis go to "Insert", "XLabel", and type in the label. See the labeled plot in Figure 3-2-6-3.

56

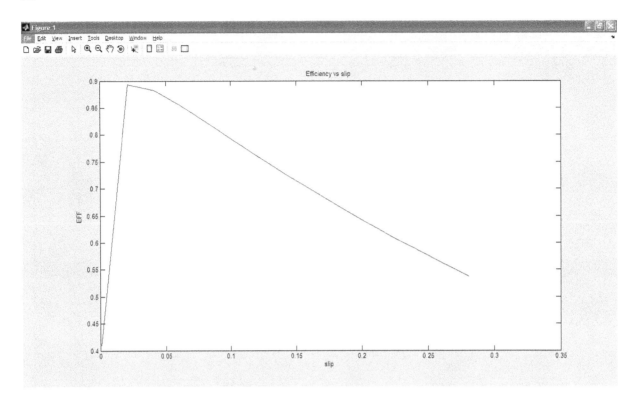

Figure 3-2-6-3 Labeled plot of an AC induction motor's "efficiency" versus "slip".

7) A smoother plot could be produced by calling for more array points.

8) There are a great number of options in MATLAB plotting. These can be accessed in the menus in the "Figure" window or by typing "plottools" at the "Command Window" prompt.

3.2.7   AC INDUCTION MOTOR 3-D PLOT OF EFFICIENCY VERSUS SUPPLY FREQUENCY AND SPEED

3-D graphics capabilities are considered to be one of MATLAB's strengths. MATLAB even uses a 3-D graphic as its icon.

3-D plots MATLAB can produce are:
        1) Five types of line graphs (i.e. 3-D line plots)
        2) Seven types of bar graphs (i.e. mesh plots)
        3) Six types of area graphs (i.e. 3-D pie charts)
        4) Five types of surface graphs
        5) Three types of direction graphs
        6) Five types of volumetric graphs (i.e. 3-D scatter plots)

Within each type of 3-D plot there are many options, such as color, lighting, rotation, title, axis labeling, etc.

The basic set up is the same for all the 3-D plots. Three identically sized data matrices with corresponding data are created. The data for these matrices can be typed in, loaded in, or generated by equations. Then the data is displayed in a 3-D plot with one of the plot methods.

A plotted function could be a simple two dimensional equation. An example would be $f(x,y) = x^2 + y^2$ where x represents the x axis, y the y axis, and $f(x,y)$ the z axis. The plotted function can also be part of a system of equations that results in a z axis value. An example would be $f(u,v) = u^2 + v$, $u = x$, and $v = y^2$.

In this example data from the solution of a system of equations is used to create a surface plot.

Problem:
Use MATLAB to create a 3-D plot showing the efficiency of a motor producing a constant 5 hp output as its supply frequency and speed are varied. Use the same motor equivalent circuit and data as in Section 3.2.5.

MATLAB's "meshgrid" mesh creation function and "surf" 3-D surface plot function are demonstrated in this example.

58

Solution:

    1) Definitions:

        Motor output power (hp), POH

        AC frequency (Hz), f

        Motor speed (rpm), N

        Motor equivalent resistances, see Section 3-2-5 (Ω), R1, R2, and RM

    2) The equations needed are:

        Motor output power (watts), $PO = 746 \cdot POH$

        Synchronous motor speed (4 pole motor) (rpm), $NSY = 30 \cdot f$

        Motor slip (pu), $slip = 1 - N/NSY$

        Line-to-line voltage (volts), $VS = 8 \cdot f$ <Note: To avoid magnetically saturating motor laminations, motor variable frequency drives often maintain a constant voltage to frequency ratio. A more thorough investigation would vary this ratio.>

        Motor current (amps), $I = |\{PO(slip)/[3 \cdot R2 \cdot (1 - slip)]\}^{.5}|$

        Motor power loss (watts), $PL = 3 \cdot [VS^2/(3 \cdot RM) + (R1 + R2) \cdot I^2]$

        Efficiency (pu) $EFF = PO/(PO + PL)$

    3) Enter the equations into MATLAB as shown in Figure 3-2-7-1.

```
% Figure 3-2-7-1.m INDUCTION MOTOR 3-D PLOT OF EFFICIENCY
% VERSUS SUPPLY FREQUENCY AND SPEED
% Defining Fixed Variables
POH=5
R1=1.6
R2=1.4
RM=980
```

Continued

```
% Equations
PO=746*POH
[f,N]=meshgrid(.1:10:71, .1:300:2101);
NSY=30.*f
slip=1-N./NSY
VS=8.*f
I=abs((PO.*slip./(3*R2.*(1-slip))).^.5)
PL=3.*(VS.^2/(3*RM)+(R1+R2).*I.^2)
EFF=PO./(PO+PL)
surf(f,N,EFF)
% PROGRAM ABOVE, OUTPUT BELOW
```

POH =

   5

R1 =

   1.6000

R2 =

   1.4000

RM =

   980

XM =

   120

X =

   6.5000

Continued

60

PO =

   3730

NSY =

1.0e+003 *

0.0030	0.3030	0.6030	0.9030	1.2030	1.5030	1.8030	2.1030
0.0030	0.3030	0.6030	0.9030	1.2030	1.5030	1.8030	2.1030
0.0030	0.3030	0.6030	0.9030	1.2030	1.5030	1.8030	2.1030
0.0030	0.3030	0.6030	0.9030	1.2030	1.5030	1.8030	2.1030
0.0030	0.3030	0.6030	0.9030	1.2030	1.5030	1.8030	2.1030
0.0030	0.3030	0.6030	0.9030	1.2030	1.5030	1.8030	2.1030
0.0030	0.3030	0.6030	0.9030	1.2030	1.5030	1.8030	2.1030
0.0030	0.3030	0.6030	0.9030	1.2030	1.5030	1.8030	2.1030

slip =

0.9667	0.9997	0.9998	0.9999	0.9999	0.9999	0.9999	1.0000
-99.0333	0.0096	0.5023	0.6677	0.7505	0.8003	0.8336	0.8573
-199.0333	-0.9805	0.0048	0.3354	0.5012	0.6007	0.6672	0.7146
-299.0333	-1.9706	-0.4927	0.0032	0.2518	0.4011	0.5008	0.5720
-399.0333	-2.9607	-0.9902	-0.3290	0.0024	0.2015	0.3344	0.4293
-499.0333	-3.9508	-1.4877	-0.6612	-0.2470	0.0019	0.1680	0.2867
-599.0333	-4.9409	-1.9852	-0.9935	-0.4963	-0.1977	0.0016	0.1440
-699.0333	-5.9310	-2.4828	-1.3257	-0.7457	-0.3973	-0.1648	0.0014

VS =

0.8000	80.8000	160.8000	240.8000	320.8000	400.8000	480.8000	560.8000
0.8000	80.8000	160.8000	240.8000	320.8000	400.8000	480.8000	560.8000
0.8000	80.8000	160.8000	240.8000	320.8000	400.8000	480.8000	560.8000
0.8000	80.8000	160.8000	240.8000	320.8000	400.8000	480.8000	560.8000
0.8000	80.8000	160.8000	240.8000	320.8000	400.8000	480.8000	560.8000
0.8000	80.8000	160.8000	240.8000	320.8000	400.8000	480.8000	560.8000
0.8000	80.8000	160.8000	240.8000	320.8000	400.8000	480.8000	560.8000
0.8000	80.8000	160.8000	240.8000	320.8000	400.8000	480.8000	560.8000

Continued

I =

1.0e+003 *

0.1605	1.6401	2.3139	2.8317	3.2685	3.6534	4.0014	4.3215
0.0297	0.0029	0.0299	0.0422	0.0517	0.0597	0.0667	0.0730
0.0297	0.0210	0.0021	0.0212	0.0299	0.0366	0.0422	0.0472
0.0298	0.0243	0.0171	0.0017	0.0173	0.0244	0.0298	0.0345
0.0298	0.0258	0.0210	0.0148	0.0015	0.0150	0.0211	0.0258
0.0298	0.0266	0.0230	0.0188	0.0133	0.0013	0.0134	0.0189
0.0298	0.0272	0.0243	0.0210	0.0172	0.0121	0.0012	0.0122
0.0298	0.0276	0.0252	0.0225	0.0195	0.0159	0.0112	0.0011

PL =

1.0e+008 *

0.0023	0.2421	0.4819	0.7217	0.9615	1.2012	1.4410	1.6808
0.0001	0.0000	0.0001	0.0002	0.0002	0.0003	0.0004	0.0005
0.0001	0.0000	0.0000	0.0000	0.0001	0.0001	0.0002	0.0002
0.0001	0.0001	0.0000	0.0000	0.0000	0.0001	0.0001	0.0001
0.0001	0.0001	0.0000	0.0000	0.0000	0.0000	0.0000	0.0001
0.0001	0.0001	0.0000	0.0000	0.0000	0.0000	0.0000	0.0000
0.0001	0.0001	0.0001	0.0000	0.0000	0.0000	0.0000	0.0000
0.0001	0.0001	0.0001	0.0000	0.0000	0.0000	0.0000	0.0000

EFF =

0.0158	0.0002	0.0001	0.0001	0.0000	0.0000	0.0000	0.0000
0.3204	0.9780	0.3155	0.1879	0.1338	0.1038	0.0848	0.0716
0.3193	0.4848	0.9829	0.4768	0.3144	0.2343	0.1866	0.1550
0.3189	0.4127	0.5833	0.9777	0.5717	0.4033	0.3113	0.2532
0.3187	0.3841	0.4823	0.6467	0.9677	0.6310	0.4673	0.3706
0.3186	0.3688	0.4370	0.5351	0.6884	0.9541	0.6685	0.5135
0.3185	0.3592	0.4112	0.4799	0.5751	0.7155	0.9375	0.6913
0.3185	0.3527	0.3945	0.4470	0.5145	0.6049	0.7319	0.9183

Figure 3-2-7-1 MATLAB program to produce data for a 3-D plot of motor efficiency versus motor speed and input voltage frequency.

4) The statement, "[f,N]=meshgrid(.1:10:71, .1:300:2101);", directs MATLAB to solve the equations following it for values of f (frequency) from .1 Hz to 71 Hz every 10 Hz and for values of N (speed) for values of .1 rpm to 2101 rpm for every 300 rpm.

5) Notice how the "./", ".*", and ".^" array arithmetic operators are used so that the equations can be solved at each of the array locations.

6) The created data is displayed in the program results for each f and N.

7) The statement, "surf(f,N,EFF)", directs MATLAB to use the created data to make a surface plot of EFF versus f and N.

8) The plot, titled "Figure 1", appeared over the "Command Window" when the program was run. If desired, the plot can be saved. It is shown in Figure 3-2-7-2.

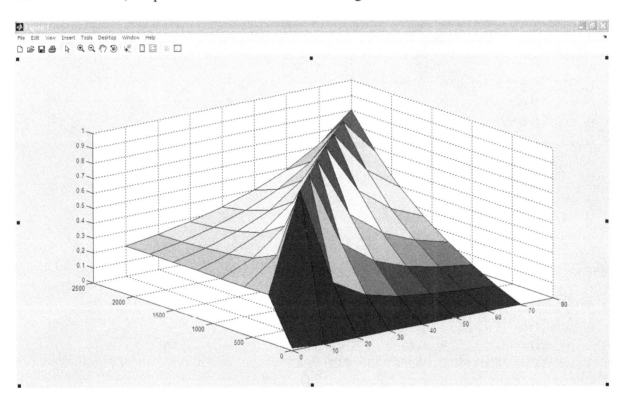

Figure 3-2-7-2 MATLAB 3-D plot of motor efficiency versus motor speed and input voltage frequency.

9) Title and axes labels can be added to the 3-D plot:

       a) In the Figures window left click on the "Insert" heading.

       b) Select X Label, Y Label, Z Label, or Title as desired.

       c) After the Labels and Title have been typed they can be moved by holding down the left mouse button and dragging them to where desired.

The plot shown in Figure 3-2-7-3 results.

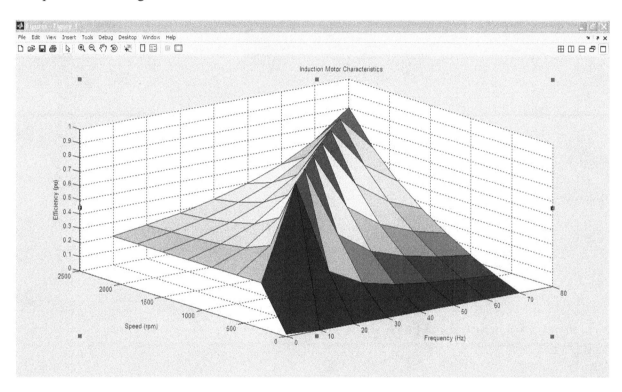

Figure 3-2-7-3 MATLAB 3-D plot of motor efficiency versus motor speed and input voltage frequency with labeled axes and a title..

3.2.8 SIMPLE STATIC DC DIODE CIRCUIT

Problem:

Use MATLAB to determine the operating voltage and current of a diode in series with a DC voltage source and resistor. The circuit is shown in Figure 3-2-8-1. Use a resistor, R, value of 5 Ω and a DC voltage supply, VS, of 1 volt. The diode characteristic curve data is supplied in a data chart.

MATLAB's "interpolation" and "2-D plotting" are demonstrated in this example.

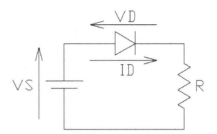

Figure 3-2-8-1 Diode in series with a DC supply and resistor.

Solution:

1) The voltage loop equation is:

$$VS = VD + ID \cdot R$$

2) Create the program of Figure 3-2-8-2.

```
% Figure 3-2-8-2.m SIMPLE STATIC DC DIODE CIRCUIT
% Input data
VS=1;
R=5;
VD= [-11.7 -11 -10.9 -10 -5 0 .7 .75 .85 1 1.2 1.5 2 2.2];
ID= [-30 -2.5 -1 -.2 -.05 0 .1 .2 .5 1 2 8 20 30];
VDi=-11.7:.05:2.2;%Voltages where current is interpolated
IDi=interp1(VD,ID,VDi);%Interpolated currents at each voltage
plot(VD,ID,'o',VDi,IDi,'x')
VDitry=-11.7;
IDitry=30;
diffmin= abs(VS - VDitry - IDitry*R);
N=((2.2+11.7)/.05)-1; %Number of interpolated data points
```

Continued

```
for n=1:N
diff = abs(VS - VDi(n) - IDi(n)*R);
if diff<diffmin
   diffmin=diff;
   VD=VDi(n);
   ID=IDi(n);
end
end
VD=VD
ID=ID
diffmin=diffmin
% Solution check
CheckSourceVoltageVS=VD + ID*R
% PROGRAM ABOVE, OUTPUT BELOW
```

VD =

   0.6000

ID =

   0.0857

diffmin =

   0.0286

CheckSourceVoltageVS =

   1.0286

Figure 3-2-8-2 MATLAB program to determine the voltage across and current through a diode.

3) Notice ";"s are used to stop some of the output values from being displayed. Using the ";"s eliminated approximately 111 pages of unneeded output.

4) The plot of the program is seen in Figure 3-2-8-3.

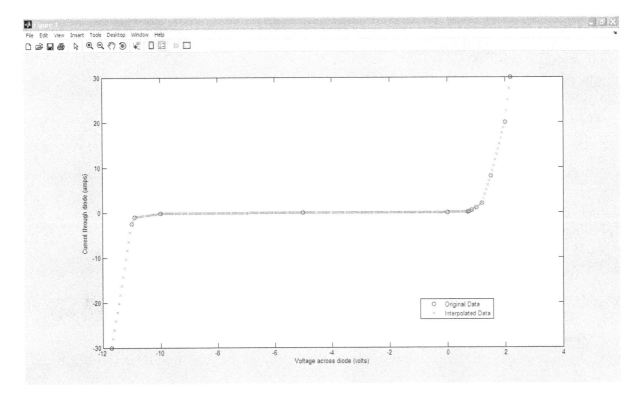

Figure 3-2-8-3 Plot of current through the diode versus the voltage across it.

5) The diode data is input in one-dimensional arrays VD and ID.

6) VDi is a one-dimensional array of voltage data spaced .05 V apart. These voltage data are values, VDi(1), VDi(2), etc. They are used as the locations to determine the interpolated currents.

7) In the "interp1" statement, VD and ID are one-dimensional arrays of the same length with related data. The "interp1" statement creates interpolated IDi values at each VDi voltage.

8) The "plot" statement makes a plot of the original current vs. voltage data with "o"s and the interpolated data with "x"s.

9) The "for" and "if" statements make the program look for interpolated voltage and current values that best solve the voltage loop equation.

3.2.9 SIMPLE STEADY-STATE AC DIODE CIRCUIT

Problem:

Use MATLAB to determine the voltage versus time across a diode in series with an AC voltage source and resistor. The circuit is shown in Figure 3-2-9-1. Use a resistor, R, value of 5 Ω and an AC voltage supply, VS, equal to 20·sin(377·t). The diode data is the same as used in Section 3.2.8.

The importance of clearing MATLAB's "Workspace" is demonstrated in this example.

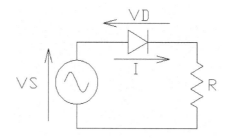

Figure 3-2-9-1 Diode in series with an AC supply and resistor.

Solution:

1) The voltage loop equation is:

$$VS = 20 \cdot \sin(377 \cdot t) = VD + ID \cdot R$$

2) Create the program of Figure 3-2-9-2.

68

```
% Figure 3-2-9-2.m SIMPLE STEADY-STATE AC DIODE CIRCUIT
% Input data
VD= [-11.7 -11 -10.9 -10 -5 0 .7 .75 .85 1 1.2 1.5 2 2.2];
ID= [-30 -2.5 -1 -.2 -.05 0 .1 .2 .5 1 2 8 20 30];
VDi=-11.7:.05:2.2;%Voltages where currents are interpolated
IDi=interp1(VD,ID,VDi);%Interpolated currents at each voltage
for k=1:100
t(k)=k/3000;
VS(k)=20*sin(377*t(k));
VDitry=-11.7;
IDitry=30;
diffmin= abs(VS(k) - VDitry - IDitry*5);
N=((2.2+11.7)/.05)-1; %Number of interpolated data points
for n=1:N
diff = abs(VS(k) - VDi(n) - IDi(n)*5);
if diff<diffmin
   diffmin=diff;
    VDf(k)=VDi(n);
end
end
end
plot(t,VDf,'o')
```

Figure 3-2-9-2 MATLAB program to plot the voltage across a diode versus time.

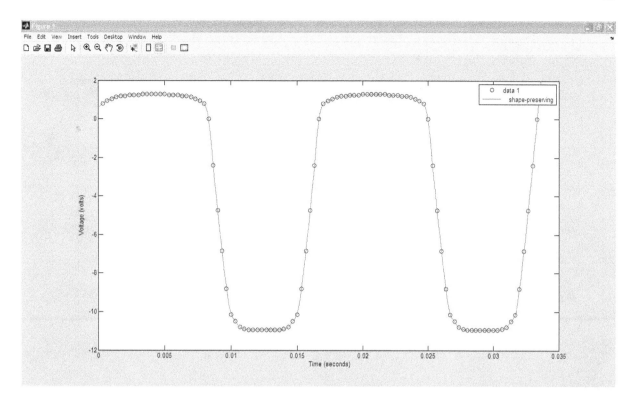

Figure 3-2-9-3 Plot of voltage across the diode versus time.

3) In Figure 3-2-9-3 the "shape-preserving interpolant" feature was used to connect the "o" data points into a relatively smooth curve. To use this feature go to the "Tools" menu, select "Basic Fitting", and select "shape-preserving interpolant".

4) This program is a good opportunity to see what happens to the plot if the step size is changed and the program rerun without clearing the "Workspace" of old values. Change the line with "for k=1:100" to "for k=1:33", run the program again, and observe the plotted output. It is shown in Figure 3-2-9-4.

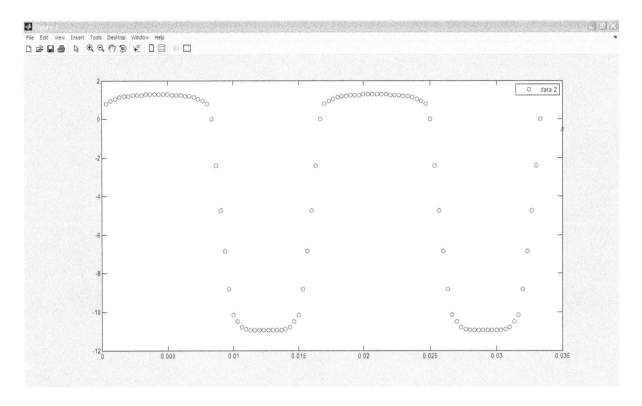

Figure 3-2-9-4 Plot of voltage across the diode versus time with K varying from 1 to 33 rather than 1 to 100 and the previous "Workspace" variable and data not cleared.

5) Notice how Figure 3-2-9-4 incorrectly appeared as the same plot as in Figure 3-2-9-3 but in Figure 3-2-9-5, where the "Workspace had been cleared", a shorter section of the plot correctly appears.

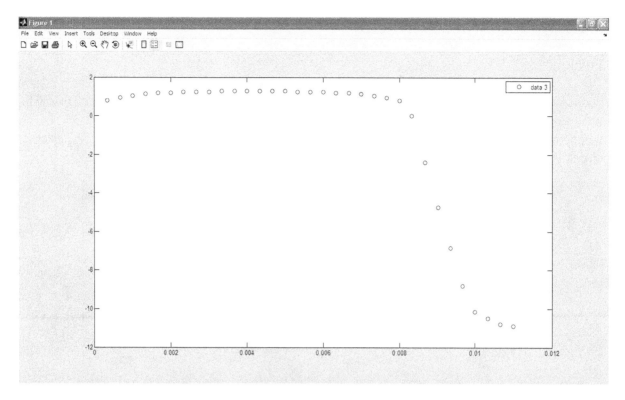

Figure 3-2-9-5 Plot of voltage across the diode versus time with K varying from 1 to 33 rather than 1 to 100 and the "Workspace" variable and data cleared.

3.2.10 INDUCTOR CHARACTERISTICS

Problem:

Flux versus current curves are known for an iron core inductor. Ignoring eddy current losses and conductor resistance losses what is the voltage across its coil and the power delivered to it when a sinusoidal AC current is forced through it? The sinusoidal current is 6 amps peak at 60 Hz. The magnetization curve, flux (Wb) versus current (amps.) is given in Figure 3-2-10-1. The electrical circuit is shown in Figure 3-2-10-2.

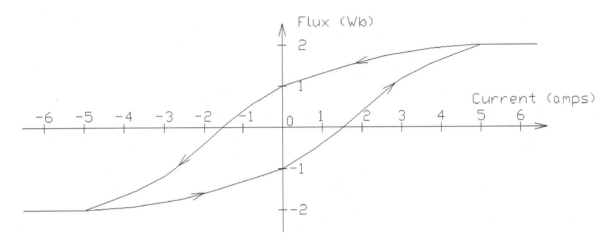

Figure 3-2-10-1 Inductor hysteresis curves. The vertical axis represents the flux enclosed by the coil (flux in the iron times the number of coil turns). The horizontal axis represents the current through the coil.

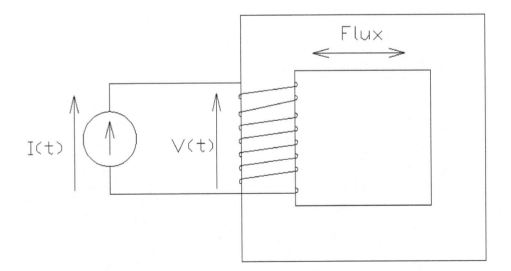

Figure 3-2-10-2 Inductor circuit. The supply is an ideal current source that forces sinusoidal 6 amp peak 60 Hz current through the inductor.

MATLAB's ability to plot two one-dimensional arrays on one plot and place several separate plots on one page with "subplot" are demonstrated in this example.

Solution:

1) The magnetization curves in Figure 3-2-10-1 apply to the condition where the current is steadily varying from less than -5 amps to greater than 5 amps. If a smaller current were applied then curves with a lesser current range would be needed. This magnetization curve does not apply to the startup condition, where the magnetization curve would start at zero flux and zero current.

2) As already stated, this problem is only concerned with the hysteresis losses. A more complete analysis of the inductor would also consider its eddy current losses and resistance losses.

3) The magnetization curves of Figure 3-2-10-1 are simplified and represented by linear equations. There are equations for the upper and lower curves. Different linear equations are used in different current ranges. The flux curve equations are in the program of Figure 3-2-10-3 and the plot they produce are in Figure 3-2-10-4.

```
% Figure 3-2-10-3.m INDUCTOR HYSTERESIS
% Demonstration of how MATLAB creates flux versus current data points
% for the hysteresis loop
% FU represents 201 points in the upper curve.
% FL represents 201 points in the lower curve.
for k=1:200
I(k)=(20/199)*k-(2010/199);
if I(k)<-5
   FU(k)=-2;
   FL(k)=-2;
end
if I(k)>=-5 & I(k)<-3
   FU(k)=.4*I(k);
   FL(k)=-1.5+.1*I(k);
end
if I(k)>=-3 & I(k)<0
   FU(k)=1+2.2*I(k)/3;
   FL(k)=-1+.8*I(k)/3;
end
if I(k)>=0 & I(k)<3
   FU(k)=1+.8*I(k)/3;
   FL(k)=-1+2.2*I(k)/3;
end
```

Continued

```
if I(k)>=3 & I(k)<5
    FU(k)=1.5+.1*I(k);
    FL(k)=.4*I(k);
end
if I(k)>=5
    FU(k)=2;
    FL(k)=2;
end
end
plot(I,FU,'*',I,FL,'.')
% PROGRAM ABOVE, OUTPUT BELOW
```

Figure 3-2-10-3 Program plotting upper and lower magnetization curves.

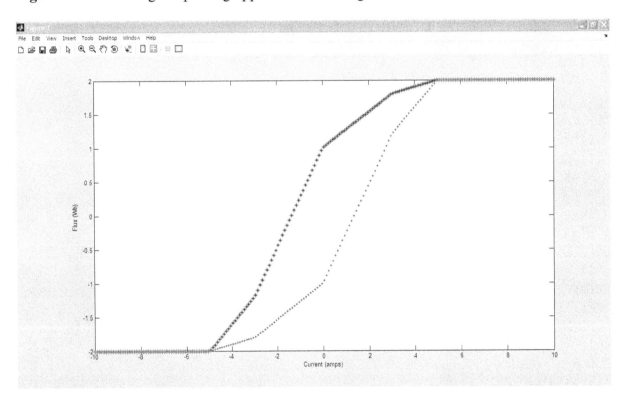

Figure 3-2-10-4 Resultant plot of the program of Figure 3-2-10-3.

4) The program of Figure 3-2-10-5 solves for the hysteresis induced voltage across the coil and the hysteresis power to the coil. The voltage across the coil is the derivative of the flux through it with respect to time. The same flux equations are used as in the program of Figure 3-2-10-3.

The program determines which flux equations to use with conditional statements. The upper or lower equations are selected by determining whether the slope of the sine wave current is increasing or decreasing. The direction of the slope is determined with the cos(377·t) value. When it is positive the current is increasing and the upper curves are used. When it is 0 the current is assumed to be increasing and the upper curves are used. When it is negative the current is decreasing and the lower curves are used.

5) The "subplot" statements break the plot page into three plots. The numbers after the parentheses configure the plots. For example, consider the second "subplot" statement, "subplot(3,1,2)". The (3,1,2) indicates that there are 3 plots stacked vertically, that there is a single column of plots, and that the "plot" statement just after it in the program is the second to be plotted.

6) The "k3=2:2000" loop effectively integrates the power over three cycles and divides by the time of three cycles to produce the average rms hysteresis power used by the coil.

```
% Figure 3-2-10-5.m INDUCTOR HYSTERESIS WITH AC APPLIED
% Flux and current data are combined with the current
% going into the coil versus time to find the flux going
% through the core versus time.
% The time span is three wavelengths.
% The k1=1:200 loop determines one-dimensional array values for time (t),
% current (I), and flux (F).
for k1=1:2000
  t(k1)=k1*3/2000/60;
  I(k1)=6*sin(377*t(k1));
%   Selecting the fluxes.
  if I(k1)<-5
    F(k1)=-2;
  end
  if I(k1)>=-5 & I(k1)<-3 & cos(377*t(k1))<0
    F(k1)=.4*I(k1);
  end
  if I(k1)>=-3 & I(k1)<0 & cos(377*t(k1))<0
    F(k1)=1+2.2*I(k1)/3;
  end
  if I(k1)>=0 & I(k1)<3 & cos(377*t(k1))<0
    F(k1)=1+.8*I(k1)/3;
  end
  if I(k1)>=3 & I(k1)<5 & cos(377*t(k1))<0
```

Continued

```
        F(k1)=1.5+.1*I(k1);
    end
    if I(k1)>=-5 & I(k1)<-3 & cos(377*t(k1))>=0
        F(k1)=-1.5+.1*I(k1);
    end
    if I(k1)>=-3 & I(k1)<0 & cos(377*t(k1))>=0
        F(k1)=-1+.8*I(k1)/3;
    end
    if I(k1)>=0 & I(k1)<3 & cos(377*t(k1))>=0
        F(k1)=-1+2.2*I(k1)/3;
    end
    if I(k1)>=3 & I(k1)<5 & cos(377*t(k1))>=0
        F(k1)=.4*I(k1);
    end
    if I(k1)>=5
        F(k1)=2;
    end
end
% The k2=2:200 loop determines one-dimensional array values for voltage (V)
% and power (P).
for k2=2:2000
    V(k2)=(F(k2)-F(k2-1))*2000*60/3;
    P(k2)=V(k2).*I(k2);
end
subplot(3,1,1)
plot(t,V,'.')
subplot(3,1,2)
plot(t,I,'.')
subplot(3,1,3)
plot(t,P,'.')
PT=0;
% The k3=2:200 loop and the equation that follows it determines
% the average rms power to the coil.
for k3=2:2000
    PT = PT+P(k3);
end
PRMSAV=PT/1999
% PROGRAM ABOVE, OUTPUT BELOW

PRMSAV =

 552.4503
```

Figure 3-2-10-5 Program plotting coil voltage, current, and power versus time.

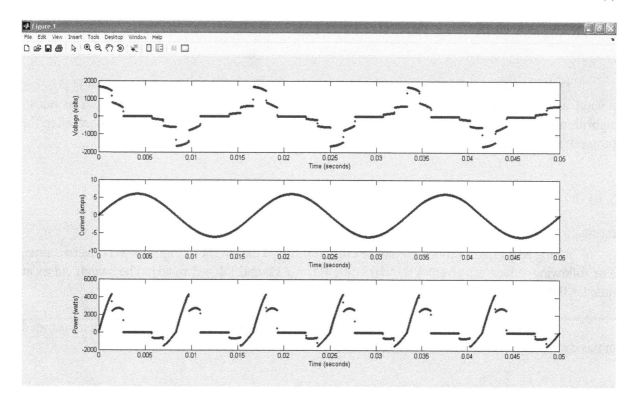

Figure 3-2-10-6 Resultant plots of the program of Figure 3-2-10-5.

3.3 TRANSIENTS IN ELECTRICAL CIRCUITS

Differential equations are used to describe transients in electrical circuits. MATLAB has several algorithms to solve a variety of differential and partial differential equations. The ode45 algorithm is shown here. Other algorithms can be used by just putting another number after the program's ode. For example, ode45 could be changed to to ode23.

3.3.1 RL CIRCUIT, FIRST ORDER DIFFERENTIAL EQUATION

Problem:
Using MATLAB, produce a plot of current through the circuit of Figure 3-3-1-1 versus time. The following values are given: VS = 10 volts, R1 = 3 Ω, and L1 = 3 milliH. The switch closes at time t = 0 seconds and the initial current is 0 A.

MATLAB's ordinary differential equation solving and plot labeling features are demonstrated in this example.

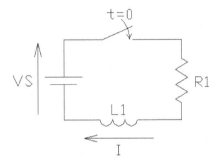

Figure 3-3-1-1 Simple inductor transient circuit.

Solution:
1) The loop equation for the circuit at times greater than zero is:

$$VS = R1{\cdot}I(t) + L1{\cdot}dI(t)/dt$$

Initial condition is I(0) = 0 A

2) MATLAB requires that the derivative be on the left hand side, so the equation above becomes:

$$dI(t)/dt = (VS - R1{\cdot}I(t))/L1$$

3) MATLAB solves differential equations in subroutines. A subroutine is written that describes the differential equation being solved. It is saved in the MATLAB default directory, so that the main program can find it when the main program is run. The subroutine is in Figure 3-3-1-2.

```
function dIdt = TRANRL(t,I)
VS=10;
R1=3;
L1=.003;
%TRANRL Defines the transient RL loop equation for Figure 3-3-1-1
dIdt = [(VS-I*R1)/L1];
```

Figure 3-3-1-2 MATLAB subroutine to determine the current through the RL circuit.

4) The main program calls for the type of differential equation solution, provides initial conditions, and a range for the solution. Details on these can be seen in the program comments. Notice that this program specifies labels for the X-axis, Y axis, and title in the program. The main program can be seen in Figure 3-3-1-3.

```
% Figure 3-3-1-3.m DIFFERENTIAL EQUATION SOLUTION OF TRANSIENT RL CIRCUIT
% ode45 is a good ordinary differential solving method for non-stiff
% systems
% @TRANRL sends the program to the default MATLAB directory that contains
% the TRANRL routine at ...MATLAB\TRANRL.m
% [0,.005] sets the evaluation time from 0 to .005 seconds
% [0] sets the initial value of I to 0 A
[t,I]=ode45(@TRANRL,[0,.005],[0]);
plot(t,I)
title('Transient RL Circuit');
xlabel('Time (seconds');
ylabel('Current (amps)');
```

Figure 3-3-1-3 MATLAB program to determine the current through the RL circuit.

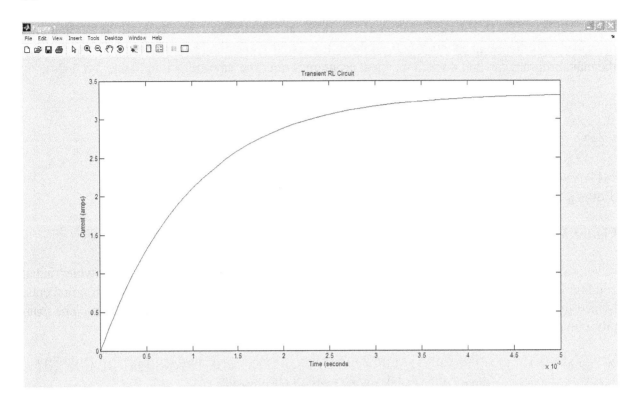

Figure 3-3-1-4 Output plot of current versus time through the RL circuit.

3.3.2 RLC CIRCUIT, SECOND ORDER DIFFERENTIAL EQUATION

To solve higher order differential equations with MATLAB the equations must first be converted to a system of first order differential equations.

For example, consider the second order differential equation:

$$dy^2/dt^2 + dy/dt + y + 1 = 0$$

Initial conditions

$$y(t = 0) = k1$$

$$dy(t = 0)/dt = k2$$

y would be made equal to y_1

dy/dt would be made equal to y_2

The equation would be rewritten as two first order differential equations with initial conditions for y_1 and y_2:

$$dy_1/dt = y_2$$

$$dy_2/dt + y_2 + y_1 + 1 = 0$$

Initial conditions

$$y_1(t = 0) = k1$$

$$y_2(t = 0) = k2$$

MATLAB would consider y_1 and y_2 as two values in a y column array.

MATLAB requires that the derivatives be on the left hand side, so the equations become:

$$dy_1/dt = y_2$$

$$dy_2/dt = -y_2 - y_1 - 1$$

82

Problem:

Use MATLAB to produce a plot of the current through the circuit of Figure 3-3-2-1 versus time. The following values are given: VS = 10 volts, R1 = 3 Ω, L1 = 3 mH, and C1 = 3 microF. The switch closes at time t = 0 seconds, the initial current is 0 amps, and the initial voltage across the capacitor is 0 volts.

MATLAB's second order differential equation solving and plotting of a one column matrix are demonstrated in this example.

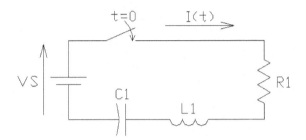

Figure 3-3-2-1 DC capacitor/inductor/resistor circuit.

Solution:

1) The loop equation for the circuit at times greater than zero is:

$$VS = R1 \cdot I(t) + L1 \cdot dI(t)/dt + (1/C1) \cdot \int I(t)dt$$

2) The equation is differentiated to put it in a form that MATLAB can solve.

$$d/dt(VS) = d/dt[R1 \cdot I(t) + L1 \cdot dI(t)/dt + \int [I(t)/C1]dt]$$

This reduces to:

$$0 = R1 \cdot dI(t)/dt + L1 \cdot d^2I(t)/dt^2 + I(t)/C1$$

The initial conditions are:

$$I(t = 0) = 0 \text{ A}$$

$$dI(t = 0^+)/dt = VS/L1 \text{ A/second}$$

3) For entry into MATLAB, the equations are converted to I(1) and I(2) where the 1 and 2 represent array subscripts:

$$dI(1)/dt = I(2)$$

$$dI(2)/dt = (-R1 \cdot I(2) - I(1)/C1)/L1$$

$$I(1) = 0 \text{ A at time } 0$$

$$I(2) = VS/L1 \text{ at time } 0^+$$

4) As in Section 3.3.1, a main program and subroutine program are required.

5) The subroutine defines the differential equation being solved. It should be saved in the MATLAB default directory, so that the main program can find it. It can be seen in Figure 3-3-2-2.

```
function dIdt = TRANRLC(t,I)
VS=10;
R1=3;
L1=.003;
C1= 3e-6;
%TRANRLC Defines the transient RLC loop equation for Figure 3-3-2-1
dIdt= [I(2);(-R1*I(2)-I(1)/C1)/L1];
```

Figure 3-3-2-2 MATLAB subroutine to determine the current through the RLC circuit.

6) The main program calls the subroutine. It supplies the initial conditions and output commands. It is shown in Figure 3-3-2-3.

84

```
% Figure 3-3-2-3.m DIFFERENTIAL EQUATION SOLUTION OF TRANSIENT RLC CIRCUIT
% ode45 is a good ordinary differential solving method for non-stiff systems
% @TRANRLC sends the program to the default MATLAB directory that contains
% the TRANRLC routine at ...MATLAB\TRANRLC.m
VS = 10;
L1 = .003;
t0 = 0; % start time of evaluation
tf = .005; % finish time of evaluation
I0 = [0 VS/L1]; % initial values of array I, 0 A and VS/L1 A/second
[t,I] = ode45(@TRANRLC,[t0,tf],I0);
plot(t,I(:,1))% the ':,' is needed to choose '1' column values
title('Transient RLC Circuit');
xlabel('Time (seconds)');
ylabel('Current (amps)');
```

Figure 3-3-2-3 MATLAB program to determine the current through the RLC circuit.

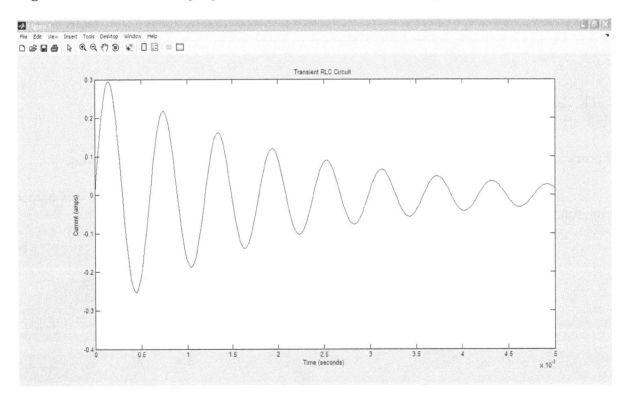

Figure 3-3-2-4 Output plot of current versus time through the RLC circuit.

3.3.3 AC INDUCTION MOTOR DRIVING A RECIPROCATING PUMP, SYSTEM OF FIRST ORDER DIFFERENTIAL EQUATIONS

A motor driving a reciprocating pump load requires transient analysis methods. The D-Q axis theory methods are the most accurate for doing transient analysis of motors. However, the D-Q axis methods are not familiar to many people and are not intuitive.

The induction motor equivalent circuit that was used in Section 3.2.5 is designed to describe an induction motor during steady-state operation. In transient conditions the equivalent circuit values may vary. Inductances and losses are not constant as a motor experiences rapid speed, voltage, and current changes. However, for the sake of presenting an example that doesn't require an advanced college degree, the less accurate, but more easily understood, equivalent circuit method will be used.

Problem:

The same 5 hp motor as in Section 3.2.5 is used to drive a reciprocating pump. The phasor equivalent circuit of Section 3.2.5 is converted to the time-based equivalent circuit of Figure 3-3-3-1. Using the same circuit values as in Section 3.2.5 plot the current, speed, and torque for .2 seconds after starting.

Pump load torque can be modeled with the equation: $T = K0 + K1 \cdot N(t) + K2 \cdot \sin\{(2 \cdot \pi \cdot [N(t)/120] \cdot t\} + K3 \cdot dN(t)/dt$. Here $K0 = .3$ ft.lb., $K1 = .01$ ft.lb./rpm, $K2 = 1.0$ ft.lb., and $K3 = .3$ ft.lb. sec^2/rpm^2.

Assume the speed of the motor at time 0, $N(0)$, is close to 0 at .0001 rpm and the initial current to the motor at time 0, $I(0)$, is 0 amps. In the program, the initial speed of the motor cannot be zero or else a divide by zero error would occur.

$f_{line} = 60$ Hz

$N_{syn} = 1800$ rpm

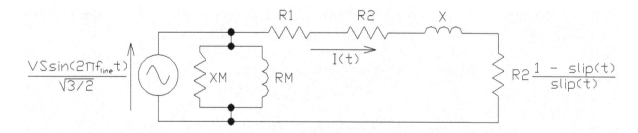

Figure 3-3-3-1 Time-based induction motor equivalent circuit of one line-to-neutral phase.

Solution:

1) See Section 3.2.5 for the equivalent circuit values.

2) The equations needed are:

Slip as a function time:

(i) $$\text{slip}(t) = [N_{syn} - N(t)]/N_{syn}$$

Motor output power as a function of time (watts):

(ii) $$PO(t) = 3 \cdot I(t)^2 \cdot R2 \cdot [1 - \text{slip}(t)]/\text{slip}(t)$$

Torque received by the load as a function of time (ft-lb):

(iii) $$T(t) = [PO(t)/746] \cdot 5252/N(t)$$

Equivalent circuit line to neutral input voltage as a function of time (volts):

(iv) $$\{VS/[SQRT(3/2)]\} \cdot \sin(2 \cdot \pi \cdot f_{line} \cdot t) = I(t) \cdot [R1 + R2/\text{slip}(t)] + [dI(t)/dt] \cdot [X/(2 \cdot \pi \cdot f_{line})]$$

Total motor input current as a function of time (amps):

(v) $$IT(t) = I(t) + \{VS/[SQRT(3/2)]\} \cdot \sin(2 \cdot \pi \cdot f_{line} \cdot t)/RM + \{VS/[SQRT(3/2)]\} \cdot \sin(2 \cdot \pi \cdot f_{line} \cdot t - \pi/2)/XM$$

Pump load torque as a function of time (watts):

(vi) $$T(t) = K0 + K1 \cdot N(t) + K2 \cdot \sin\{2 \cdot \pi \cdot [N(t)/120] \cdot t\} + K3 \cdot dN(t)/dt.$$

3) Rewrite equations (iv) and (vi) with the differentials on the left side of the equals sign to make them ready for entry into a MATLAB differential equation solving subroutine.

(iv') $$dI(t)/dt = \{VS/[SQRT(3/2)] \cdot \sin(2 \cdot \pi \cdot f_{line} \cdot t) - I(t) \cdot [R1 + R2/\text{slip}(t)]\}/[X/(2 \cdot \pi \cdot f_{line})]$$

(vi') $$dN(t)/dt = (T(t) - K0 - K1 \cdot N(t) - K2 \cdot \sin\{2 \cdot \pi \cdot [N(t)/120] \cdot t\})/K3$$

4) As in Section 3.3.2 the differential equations (iv') and (vi') are solved as separate columns of a matrix. The I(t) will be renamed as A(1). The N(t) will be renamed as A(2).

5) Other variables in the equations will also be renamed as columns of the A matrix. A(3) is motor torque, T. A(4) is total motor current, IT. A(5) is motor slip, SLIP. A(6) is motor output power, PO.

6) The equations are used in the MATLAB differential equation solving subroutine shown in Figure 3-3-3-2.

```
function DADT = MOTORPUMP(t,A)
VS=480;
FLINE=60;
R1=1.6;
R2=1.4;
RM=980;
X=6.5;
XM=120;
K0=.03;
K1=.001;
K2=2;
K3=.003;
NSYN=1800;
% A(1)is current, I(t), in amps
% A(2) is motor speed, N(t), in rpm
% A(3) is motor torque, T, in ft-lb.
% A(4) is total motor current, IT, in amps
% A(5) is motor slip, SLIP
% A(6) is motor output power, PO, in watts
A(5)=(NSYN-A(2))/NSYN;
A(6)=3*A(1)^2*R2*(1-A(5))/A(5);
A(3)=(A(6)/746)*5252/A(2);
A(4)=A(1)+(VS/(sqrt(3/2)))*sin(6.283*FLINE*t)/RM+(VS/(sqrt(3/2)))*sin(6.283*FLINE*t-1.571)/XM;
DADT = [(VS/(sqrt(3/2))*sin(6.283*FLINE*t)-A(1)*(R1+R2/A(5)))/(X/(6.283*FLINE));
(A(3)-K0-K1*A(2)-K2*sin(6.283*(A(2)/500)*t))/K3];
```

Figure 3-3-3-2 MATLAB subroutine to determine the current, speed, and torque of an induction motor driving a reciprocating pump load.

7) The main program calls the subroutine. It supplies the initial conditions and output commands. Array element values are calculated one element at a time with the use of the array operators, ".*", "./" and ".^". The proper columns are called out with the use of ":,". For example, A(:,2) calls out the elements of column 2. The main program is shown in Figure 3-3-3-3.

```
% Figure 3-3-3-3.m MOTOR AND RECIPROCATING PUMP
% ode45 is a good ordinary differential solving method for non-stiff
% systems
% @MOTORPUMP sends the program to the default MATLAB directory that contains
% the MOTORPUMP routine at ...MATLAB\MOTORPUMP.m
t0=0; % start time of evaluation
tf=.2; % finish time of evaluation
A0=[0 .0001]; % initial values of array A (current and speed)
% Call for ordinary differential equation solution
[t,A]=ode45(@MOTORPUMP,[t0,tf],A0);
% Values used in the equations
VS=480;
FLINE=60;
R2=1.4;
RM=980;
XM=120;
NSYN=1800;
% A(:,1)is current, I(t), in amps
% A(:,2) is motor speed, N(t), in rpm
% A(:,3) is motor torque, T, in ft-lb.
% A(:,4) is total motor current, IT, in amps
% A(:,5) is motor slip, SLIP
% A(:,6) is motor output power, PO, in watts
%Calculating values at each array element value
A(:,5)=(NSYN-A(:,2))/NSYN;
A(:,6)=3.*A(:,1).^2.*R2.*(1- A(:,5))./A(:,5);
A(:,3)=(A(:,6)./746).*5252./A(:,2);
A(:,4)=A(:,1)+(VS/(sqrt(3/2)))*sin(6.283*FLINE*t)/RM+(VS/(sqrt(3/2)))*sin(6.283*FLINE*t-
1.571)/XM;
%Output plotting
subplot(3,1,1)
plot(t,A(:,4))
title('Motor Input Current');
xlabel('Time (seconds)');
ylabel('Current (amps)');
subplot(3,1,2)
plot(t,A(:,3))
title('Motor Torque');
xlabel('Time (seconds)');
ylabel('Torque (ft-lb.)');
subplot(3,1,3)
```

Continued

```
plot(t,A(:,2))
title('Motor Speed');
xlabel('Time (seconds)');
ylabel('Speed (rpm)');
```

Figure 3-3-3-3 MATLAB program to determine the current, speed, and torque of an induction motor driving a reciprocating pump load.

8) The plotted outputs are shown in Figure 3-3-3-4.

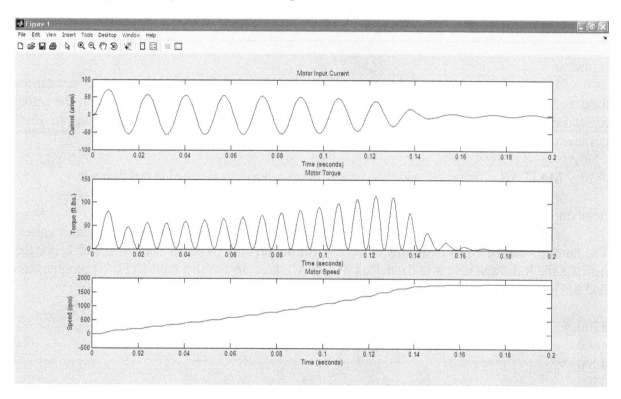

Figure 3-3-3-4 Output plot of current, speed, and torque of an induction motor driving a reciprocating pump load.

3.4 MISCELLANEOUS EXAMPLES

3.4.1 STATISTICAL ANALYSIS OF RESISTOR RESISTANCES

MATLAB can do basic statistical analysis. It can determine or calculate maximum and minimum values, average values, median values, most frequent values (modes), standard deviations, variances, and make histograms. For more advanced statistical analysis, MATLAB needs its add-on Statistics Toolbox.

MATLAB can do a statistical analysis of matrices with multiple columns, although here only a one column matrix will be demonstrated.

Problem:

Twenty-five resistors of the same 4700 Ω nominal value have their actual resistances measured. Tabulate the actual resistance measurements in a MATLAB data file. Using the actual resistance measurements determine the minimum, maximum, average, median, mode, and standard deviation; then plot a histogram.

MATLAB's "statistics" and "histogram" features are demonstrated in this example.

Solution:

1) Using the MATLAB "Editor" create a 1 x 25 column matrix of actual resistance values. The numbers can be typed in directly or copied in from another program, such as Excel. Name the matrix's file R.dat and save it in the default MATLAB directory, so that the MATLAB program can find it. The matrix is in Figure 3-4-1-1.

4700
4800
4900
4700
4500
4400
5000
4300
5100
4700
4800
4600

Continued

4900
4500
5000
4700
4600
4600
4600
4600
4500
4900
4400
5100
4700

Figure 3-4-1-1 Matrix of measured resistor resistances.

2) Write the program shown in Figure 3-4-1-2.

```
% Figure 3-4-1-2.m STATISTICAL ANALYSIS OF RESISTOR RESISTANCES
% Input data
load R.dat
% Find the minimum value
minimum = min(R)
% Find the maximum value
maximum = max(R)
% Find the most common value (mode value)
mod = mode(R)
% Find the median value
med = median(R)
% Calculate the mean (average)
average = mean(R)
% Calculate the standard deviation
std_dev = std(R)
% Make a histogram of the R data with 10 evenly spaced bins
hist(R)
%Output plotting
title('Resistor Resistance Histogram');
xlabel('Resistance (ohms)');
ylabel('Number of Resistors');
% PROGRAM ABOVE, OUTPUT BELOW
```

Continued

minimum =

    4300

maximum =

    5100

mod =

    4600

med =

    4700

average =

    4704

std_dev =

  218.8607

Figure 3-4-1-2 Program to statistically analyze resistance data.

    3) The histogram statement, "hist", automatically separates the data into 10 evenly spaced bins. MATLAB has other histogram functions that allow variable width bins and different numbers of bins. The 10 bin histogram of this data is shown in Figure 3-4-1-3.

93

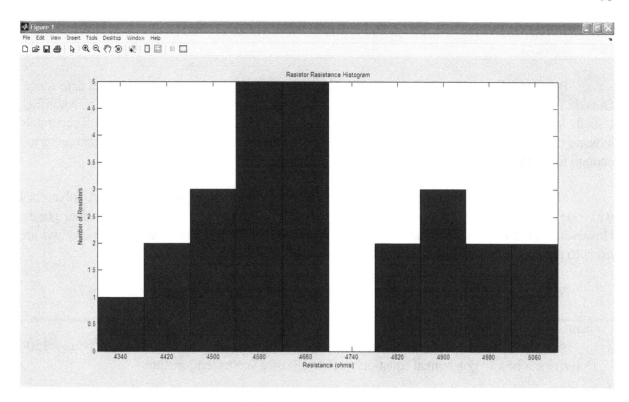

Figure 3-4-1-3 Histogram of resistor values.

3.4.2 APPROXIMATING RESISTANCE VERSUS TEMPERATURE DATA WITH A POLYNOMIAL

Polynomials can be useful for representing non-periodic data over their supplied data range. Outside of the data range they may not be very accurate, especially the higher order polynomials. Usually lower order polynomials are the most useful. A simple first order polynomial will accurately indicate the trend of data. Second and third order polynomials produce smooth curves that are good enough for many applications.

Higher order polynomials come closer to exactly crossing through data points. A polynomial with one less order than the number of its data points will exactly cross through its data points. However, higher order polynomials can produce unreasonable results between data points and are likely to produce incorrect results out of the data range.

MATLAB uses the least-squares method to determine polynomial coefficients.

Problem:

A 10 kΩ resistor has its resistance measured at different temperatures from -50° C to +150° C. Determine a best fit polynomial equation for its resistance versus temperature.

MATLAB's "polyfit", "polyval", "format", and "format short e" statements are demonstrated in this example.

Solution:

1) Create a 2 x 21 matrix of resistor temperature and resistance data. Name the matrix TR.dat and save it in the default MATLAB directory. It is shown in Figure 3-4-2-1.

-50	10.32
-40	10.27
-30	10.2
-20	10.13
-10	10.09
0	10.04
10	10.02
20	10.01
30	10
40	10
50	10
60	10.01
70	10.03
80	10.05
90	10.09

Continued

100	10.14
110	10.2
120	10.29
130	10.36
140	10.44
150	10.55

Figure 3-4-2-1 Matrix of resistor temperatures and corresponding resistances.

2) Write the MATLAB program of Figure 3-4-2-2.

```
% Figure 3-4-2-2.m POLYNOMIAL REPRESENTING A RESISTOR'S RESISTANCE
%VERSUS TEMPERATURE DATA
% Input data
load TR.dat
format % 'format' makes the output numbers 5-digit floating point
p = polyfit(TR(:,1),TR(:,2),3) % 'polyfit' uses a least squares method to find
% coefficients for the polynomial
format short e % 'format short e' makes the output numbers 5-digit floating point
% with a x 10 exponent
p = p
x = (TR(1,1):.1:TR(21,1)); % selects 2000 x values for temperature
R = polyval(p,x); % 'polyval' evaluates a polynomial with p coefficients at values x
%Output plotting
plot(TR(:,1),TR(:,2),'*',x,R)% plots data points and the polynomial
title('Resistance versus Temperature');
xlabel('Temperature (deg. C)');
ylabel('Resistance (ohms)');
% PROGRAM ABOVE, OUTPUT BELOW

p =

   0.0000   0.0000  -0.0034  10.0566


p =

 2.5060e-008 4.0596e-005 -3.3917e-003  1.0057e+001
```

Figure 3-4-2-2 Program to create a third order polynomial that represents resistor resistance versus temperature.

3) The "format" statement directs the program to output 5-digit numbers without x 10 exponents. The first output of p shows this. In this output the values of the third order and second order coefficients appear to be zero. This is misleading. The "format short e" statement outputs the same coefficients with 5 figures and a x 10 exponent. The second printing of p shows non-zero values for the same coefficients.

4) The "polyfit" statement uses the data from TR.dat to solve for the coefficients of a third order polynomial. The 3 in the statement indicates coefficients are being solved for a third order polynomial.

5) The polynomial written out with T for temperature and R for resistance is:

$R = 2.5060 \times 10^{-8} \cdot T^3 + 4.0596 \times 10^{-5} \cdot T^2 - 3.3917 \times 10^{-3} \cdot T + 1.0057 \times 10^{1}$

6) The "polyval" statement evaluates the polynomial using the p coefficients.

7) A plot of the original data and the polynomial is shown in Figure 3-4-2-3.

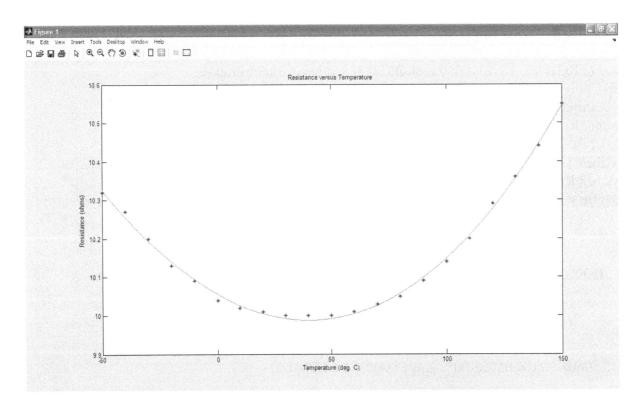

Figure 3-4-2-3 Plot of resistance versus temperature data and polynomial.

3.4.3 DETERMINING THE FREQUENCY CONTENT OF A WAVEFORM WITH THE DISCRETE FOURIER TRANSFORM

Fourier Transforms are useful to the electrical engineer for determining the frequency spectrum of non-sinusoidal periodic waveforms.

MATLAB computes the Discrete Fourier Transform using Fast Fourier Transform algorithms. The statement used in MATLAB programs is "fft" even though it actually does a Discrete Fourier Transform.

With the MATLAB Discrete Fourier Transform:
1) Data should be uniformly-spaced.
2) Sampling must be done at least at the Nyquist frequency to avoid aliasing errors. The Nyquist frequency is twice the frequency of the highest frequency component of the waveform. Generally, for good results, the sampling frequency should be at least six times the highest frequency component. More samples are better than fewer.
3) At least 3 periods should be sampled. Fractional periods may be included in the sampling. Generally, better results are obtained when more periods are sampled.
4) Fast Fourier Transforms require that the number of samples be exactly 2^n where n is some positive integer. Discrete Fourier Transform analysis, the type done by MATLAB, does not require this.
5) The Discrete Fourier Transform produces two groups of values. The values are complex conjugates of each other. Those of the higher frequency group (sometimes called the negative frequency group) are usually disregarded.

Many demonstrations of Fourier Transforms analyze a waveform equation. However, actual electrical data is not usually received in equation form. It usually appears in a graphical form or as numeric data. Here the Discrete Fourier Transform will be demonstrated with numeric data that was generated by an equation. This demonstration shows that the Fourier Transform can be used with data. After the Fourier Transform is done, its Fourier Transform amplitudes and frequencies can be compared with the frequencies and amplitudes of the original equation.

Problem:
Determine the Discrete Fourier Transform frequency spectrum for the data derived from the equations:

$$V(t) = \sin(w_0 \cdot t) + 2 \cdot \sin(2 \cdot w_0 \cdot t + .7854) + 3 \cdot \sin(3 \cdot w_0 \cdot t)$$

$$w_0 = 2 \cdot \pi \cdot f$$

$$f = 60 \text{ Hz}$$

MATLAB's "Discrete Fourier Transform" statement and a matrix column removal feature are demonstrated in this example.

Solution:

      1) Enter and run the program shown in Figure 3-4-3-1. In this run of the program the number of samples is deliberately set low, at 8, to demonstrate the program's numeric output without generating too much output data. With n set this low the sampling frequency is much less than the Nyquist frequency, so the numeric and plotted output are not realistic and are only academic.

```
% Figure 3-4-3-1.m DISCRETE FOURIER TRANSFORM

% Plotting the waveform
f = 60;
wo = 2*3.1416*f; % angular frequency of waveform
TM = .06; % time of measurement
t = (0:TM/199:TM); % selects 200 time values for plotting
VWAVE= sin(wo*t) + 2*sin(2*wo*t + .7854) + 3*sin(3*wo*t);
subplot(3,1,1)
plot(t,VWAVE) % plots smooth wave
title('Waveform');
xlabel('Time (seconds)');
ylabel('Voltage (volts)');


% Finding data points and plotting them
n = 8 % number of samples
t1 = (0:TM/(n - 1):TM); % selects time values for n data points
VDATA= sin(wo*t1) + 2*sin(2*wo*t1 + .7854) + 3*sin(3*wo*t1); % produces n
%data points
subplot(3,1,2)
plot(t1,VDATA,'*') % plots data
title('Data points used by FFT');
xlabel('Time (seconds)');
ylabel('Voltage (volts)');
```

Continued

```
% Finding the Discrete Fourier Transform
fSAMP = (n - 1)/TM; % Sampling frequency in Hz
VDFT = fft(VDATA)
VDFTM = abs(VDFT)
fl = (0:fSAMP/(n - 1):fSAMP) % Frequencies
subplot(3,1,3)
plot(fl,VDFTM)
title('Fourier Transform Amplitudes');
xlabel('Frequency (Hz)');
ylabel('Amplitude');
% PROGRAM ABOVE, OUTPUT BELOW
```

n =

   8

VDFT =

 Columns 1 through 4

  12.5974   -2.8931 - 0.4403i -2.5607 - 1.5389i -4.2561 - 4.2143i

 Columns 5 through 8

  18.1360   -4.2561 + 4.2143i -2.5607 + 1.5389i -2.8931 + 0.4403i

VDFTM =

  12.5974   2.9264   2.9876   5.9895   18.1360   5.9895   2.9876   2.9264

fl =

    0   16.6667   33.3333   50.0000   66.6667   83.3333 100.0000 116.6667

Figure 3-4-3-1 Program to demonstrate the MATLAB Discrete Fourier Analysis function with the sampling frequency set below the minimum Nyquist frequency. 8 samples are used.

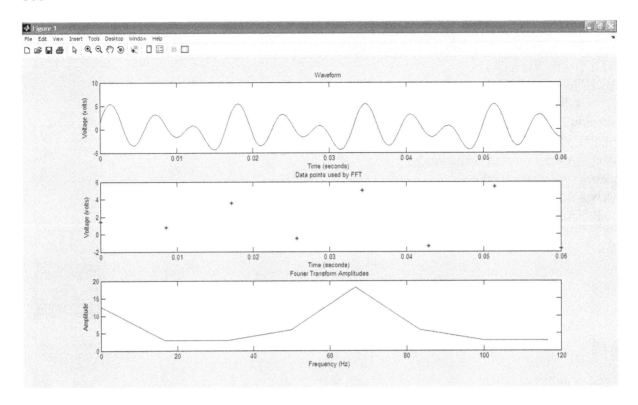

Figure 3-4-3-2 Output plot of the program of Figure 3-4-3-1.

2) The first two sections of the program setup and plot the waveform and selected 8 data points. Looking at the plots of them in Figure 3-4-2-2 it is clear that there are not enough data points to describe the waveform.

3) The third section of the program evaluates the Discrete Fourier Transform and plots it. VDFT is the numeric output of the "fft" function. The first term in VDFT is the sum of all of the V(t) data. The next four VDFT complex numbers are magnitudes of the Fourier Transform at the second through fifth frequencies listed after f1. The last three VFDT complex numbers are magnitudes of the Fourier Transform at higher sampling frequencies. These are complex conjugates to the second to fifth values. These three values are usually not used.

4) VDFTMs are the magnitudes of the VDFTs.

5) The program is rewritten slightly in Figure 3-4-3-3.

      a) More samples are taken. Often when working with data, one cannot easily determine the number of samples needed. It can take trial and error. However, in this example the equation shows that the highest component frequency is 3 x 60 = 180 Hz. Setting the sampling frequency to six times that produces 6 x 180 = 1080. Given a time of measurement of .060 seconds, n = fSAMP*TM + 1 = 65.8. This is rounded up to n = 66 for the program.

      b) Semicolons have been placed after variables that do not need to be printed in the output.

      c) "fl(:,17:66) = []" and "VDFTM(:,17:66) = []" statements have been added to cause the program to disregard the higher frequencies when plotting. This results in a clearer display of the lower frequency values.

```
% Figure 3-4-3-3.m DISCRETE FOURIER TRANSFORM

% Plotting the waveform
f = 60;
wo = 2*3.1416*f; % angular frequency of waveform
TM = .06; % time of measurement
t = (0:TM/199:TM); % selects 200 time values for plotting
VWAVE= sin(wo*t) + 2*sin(2*wo*t + .7854) + 3*sin(3*wo*t);
subplot(3,1,1)
plot(t,VWAVE) % plots smooth wave
title('Waveform');
xlabel('Time (seconds)');
ylabel('Voltage (volts)');

% Finding data points and plotting them
n = 66 % number of samples
t1 = (0:TM/(n - 1):TM); % selects time values for n data points
VDATA= sin(wo*t1) + 2*sin(2*wo*t1 + .7854) + 3*sin(3*wo*t1); % produces n
%data points
subplot(3,1,2)
plot(t1,VDATA,'*') % plots data
title('Data points used by FFT');
xlabel('Time (seconds)');
ylabel('Voltage (volts)');
```

Continued

```
% Finding the Discrete Fourier Transform
fSAMP = (n - 1)/TM; % Sampling frequency in Hz
VDFT = fft(VDATA);
VDFTM = abs(VDFT);
f1 = (0:fSAMP/(n - 1):fSAMP); % Frequencies
f1(:,17:66) = []; % deletes the rightmost 50 columns
VDFTM(:,17:66) = []; % deletes the rightmost 50 columns
subplot(3,1,3)
plot(f1,VDFTM)
title('Fourier Transform Amplitudes');
xlabel('Frequency (Hz)');
ylabel('Amplitude');
% PROGRAM ABOVE, OUTPUT BELOW

n =

 66
```

Figure 3-4-3-3 Program to demonstrate the MATLAB Discrete Fourier Analysis function with the sampling frequency above the minimum Nyquist frequency. 66 samples are used.

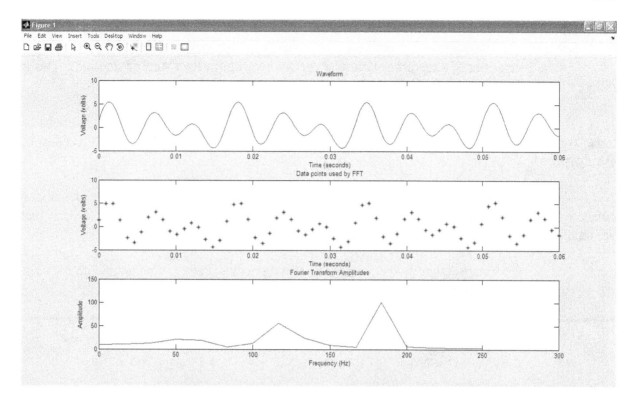

Figure 3-4-3-4 Output plot of the program of Figure 3-4-3-3.

6) Notice how the Fourier Transform magnitudes peak at approximately 60, 120, and 180 Hz, the frequencies of the original equation, and they peak at magnitudes proportional to those of the original equation.

3.4.4 FOURIER SERIES ANALYSIS

A Fourier Series is a representation of a periodic waveform by a sum of sinusoids. This is different from a Fourier Transform which produces an amplitude versus frequency spectrum.

Unlike the Fourier Transform, the finding of Fourier Coefficients requires computations be done over an integral number of periods. Usually one period is selected. Samples do not need to be evenly spaced.

With a Fourier Series a periodic waveform, f(t), can be represented as a sum of sine and cosine waves or the sum of sine waves with differing phase angles. The waves have frequencies that are multiples of a primary frequency. The Fourier Series representation of a wave is:

$$f(t) = A_0 + A_1 \cdot \sin(w_0 \cdot t) + A_2 \cdot \sin(2 \cdot w_0 \cdot t) + A_3 \cdot \sin(3 \cdot w_0 \cdot t) + ... + A_n \cdot \sin(n \cdot w_0 \cdot t)$$
$$+ B_1 \cdot \cos(w_0 \cdot t) + B_2 \cdot \cos(2 \cdot w_0 \cdot t) + B_3 \cdot \cos(3 \cdot w_0 \cdot t) + ... + B_n \cdot \cos(n \cdot w_0 \cdot t)$$

or

$$f(t) = A_0 + C_1 \cdot \sin(w_0 \cdot t + \angle\theta_1) + C_2 \cdot \sin(2 \cdot w_0 \cdot t + \angle\theta_2) + C_3 \cdot \sin(3 \cdot w_0 \cdot t + \angle\theta_3) + ... + C_n \cdot \sin(n \cdot w_0 \cdot t + \angle\theta_n)$$

The function of time, f(t), describes the waveform amplitude. "w_o" is the angular frequency of the base harmonic. "n" is the order of each harmonic. Fourier coefficients are found with the following equations:

$$A_0 = (1/T) \cdot \int_0^T f(t)\, dt$$

$$A_n = (2/T) \cdot \int_0^T f(t) \cdot \sin(n \cdot w_0 \cdot t)\, dt$$

$$B_n = (2/T) \cdot \int_0^T f(t) \cdot \cos(n \cdot w_0 \cdot t)\, dt$$

$$C_n = |A_n + B_n \cdot j|$$

$$\theta_n = \arg(A_n + B_n \cdot j)$$

T is the period of the primary harmonic frequency.

Problem:

      Determine the Fourier Series equation that represents the waveform of the data of Section 3.4.3. As in Section 3.4.3, the primary harmonic frequency is 60 Hz.

      MATLAB's "concatenation" feature and ability to do simple numerical integration are demonstrated in this example.

Solution:

      1) Determine the values of the waveform of the equation every .001 seconds over one period. The last value found should be found where the waveform starts to repeat at 1/60 seconds.
      Repeating the equation:

$$V(t) = \sin(w_0 \cdot t) + 2 \cdot \sin(2 \cdot w_0 \cdot t + .7854) + 3 \cdot \sin(3 \cdot w_0 \cdot t)$$

$$w_0 = 2 \cdot \pi \cdot f$$

$$f = 60 \ Hz$$

      2) The MATLAB program in Figure 3-4-4-1 determines V(t) for values t = 0, .001, .002, .003, .004, .005, .006, .007, .008, .009, .010, .011, .012, .013, .014, .015, .016, and .0167.

```
% Figure 3-4-4-1.m FOURIER SERIES DATA GENERATION
f = 60
wo = 2*pi*f
t = [0 .001 .002 .003 .004 .005 .006 .007 .008 .009 .010 .011 .012 .013 .014 .015 .016 .0167]
Vt = sin(wo*t) + 2*sin(2*wo*t + .7854) + 3*sin(3*wo*t)
% PROGRAM ABOVE, OUTPUT BELOW

f =

  60


wo =

  376.9911
```

Continued

t =

 Columns 1 through 9

     0   0.0010   0.0020   0.0030   0.0040   0.0050   0.0060   0.0070   0.0080

 Columns 10 through 18

  0.0090   0.0100   0.0110   0.0120   0.0130   0.0140   0.0150   0.0160   0.0167

Vt =

 Columns 1 through 9

  1.4142   5.0816   4.4963   0.3470   -3.1747   -2.7877   0.5616   3.0396   2.2478

 Columns 10 through 18

  -0.3817   -1.6589   -0.5429   0.7564   -0.2848   -3.1021   -4.3489   -1.7443   1.5749

Figure 3-4-4-1 Program to find V(t) values for values of t = 0 to .0167.

  3) The data from the program of Figure 3-4-4-1 is blocked, copied, and pasted into the program shown in Figure 3-4-4-2.


```
% Figure 3-4-4-2.m FOURIER SERIES COEFFICIENT SOLUTION
% Initial data
f = 60;
wo = 2*pi*f;
t1 = [0 0.0010 0.0020 0.0030 0.0040 0.0050 0.0060 0.0070 0.0080];
t2 = [0.0090 0.0100 0.0110 0.0120 0.0130 0.0140 0.0150 0.0160 0.0167];
Vtd1 = [1.4142 5.0816 4.4963 0.3470 -3.1747 -2.7877 0.5616 3.0396 2.2478];
Vtd2 = [-0.3817 -1.6589 -0.5429 0.7564 -0.2848 -3.1021 -4.3489 -1.7443 1.5749];
% t and Vtd are formed by concatenating t1 & t2 and Vtd1 & Vtd2
t = [t1 t2];
Vtd = [Vtd1 Vtd2];
N = 100; % Number of interpolated points
T = .0167;% Time period of the fundamental frequency
```

Continued

```
% Sets all initial values to zero
A0 = 0;
A1 = 0;
A2 = 0;
A3 = 0;
A4 = 0;
A5 = 0;
B0 = 0;
B1 = 0;
B2 = 0;
B3 = 0;
B4 = 0;
B5 = 0;
ti = 0:T/N:T; % Range of interpolated times
Vti = interp1(t,Vtd,ti); % Creates interpolated values between the data values
for k = 2:N+1 % for/end loop solves for coefficient magnitudes and angles with numeric
integration
A0 = A0 + Vti(k)/N;
A1 = A1 + 2*Vti(k)*sin(wo*ti(k))/N;
A2 = A2 + 2*Vti(k)*sin(2*wo*ti(k))/N;
A3 = A3 + 2*Vti(k)*sin(3*wo*ti(k))/N;
A4 = A4 + 2*Vti(k)*sin(4*wo*ti(k))/N;
A5 = A5 + 2*Vti(k)*sin(5*wo*ti(k))/N;
B1 = B1 + 2*Vti(k)*cos(wo*ti(k))/N;
B2 = B2 + 2*Vti(k)*cos(2*wo*ti(k))/N;
B3 = B3 + 2*Vti(k)*cos(3*wo*ti(k))/N;
B4 = B4 + 2*Vti(k)*cos(4*wo*ti(k))/N;
B5 = B5 + 2*Vti(k)*cos(5*wo*ti(k))/N;
end
% Determines output values
A0 = A0
mC1 = abs(A1 + B1*(j))
aTH1 = angle(A1 + B1*(j))
mC2 = abs(A2 + B2*(j))
aTH2 = angle(A2 + B2*(j))
mC3 = abs(A3 + B3*(j))
aTH3 = angle(A3 + B3*(j))
mC4 = abs(A4 + B4*(j))
aTH4 = angle(A4 + B4*(j))
mC5 = abs(A5 + B5*(j))
aTH5 = angle(A5 + B5*(j))
```

Continued

```
% for/end loop plots input data (Vtd), interpolated values (Vti),
% original equation (Vto), and computed Fourier Series equation (Vtfs)
for k1 = 1:N
subplot(2,1,1)
plot(t,Vtd,'o',ti,Vti)
title('Voltage Data and Interpolated Voltage Data');
xlabel('Time (seconds)');
ylabel('Voltage (volts)');
subplot(2,1,2)
Vto = sin(wo*ti) + 2*sin(2*wo*ti + .7854) + 3*sin(3*wo*ti);
Vtfs1 = A0 + mC1*sin(wo*ti + aTH1) + mC2*sin(2*wo*ti + aTH2);
Vtfs2 = mC3*sin(3*wo*ti + aTH3) + mC4*sin(4*wo*ti + aTH4) + mC5*sin(5*wo*ti + aTH5);
Vtfs = Vtfs1 + Vtfs2;
plot(ti,Vto,ti,Vtfs)
title('Original Equation and Fourier Series Equation');
xlabel('Time (seconds)');
ylabel('Voltage (volts)');
end
% PROGRAM ABOVE, OUTPUT BELOW

A0 =

   0.0029


mC1 =

   0.9865


aTH1 =

   0.0059


mC2 =

   1.9091
```

Continued

aTH2 =

0.7874

mC3 =

2.6931

aTH3 =

0.0015

mC4 =

0.0031

aTH4 =

1.1244

mC5 =

0.0034

aTH5 =

1.1127

Figure 3-4-4-2 Program to determine Fourier Series coefficients from waveform data.

4) Arrays for time, t, and voltage, Vtd, data are separated into two data sets each, t1 & t2 and Vtd1 & Vtd2. These separated data sets are concatenated into t and Vtd. Here, the reason for this is to demonstrate concatenation and to make the data easier to read. In another program the data might be more than could be put on a single program line. Then, concatenation or a separate data matrix, like that used in Section 3.4.2, must be used.

5) The 18 data points are not all evenly spaced. Most have a .001 second spacing, but the last has a .0007 second spacing. The "interp1" statement is used to linearly interpolate between the 18 data points to produce 101 evenly spaced data points. This may improve the accuracy of the final Fourier Series coefficients. However, the improvement cannot be depended on. The reason interpolation is used comes later in the program when simple numerical integration is done. The program for simple integration is easier to write with evenly spaced data.

6) In calculus integration over time, infinitesimally small changes in time are multiplied by a function at different times over the time range of the integration. Then the products are summed. An example is the equations for solving for A0:

$$A_0 = (1/T)\cdot\int_0^T f(t)\ dt$$

In the simple numerical integration method of this program the differential time, dt, is represented by the spacing between the interpolated values, T/N. The f(t) is the interpolated voltage value, V(ti), at each interpolated time value, ti. The summation is accomplished in the program by the for/end loop that adds the last value determined for A0 to the current (1/T) Vdt(ti) (T/N). This simplifies to A0 = A0 + Vdt(ti)/N.

7) The plots in Figure 3-4-4-3 show the original input data and linearly interpolated data on the upper plot. The lower plot is of the original voltage equation and of the equation made up with the computed Fourier Series coefficients.

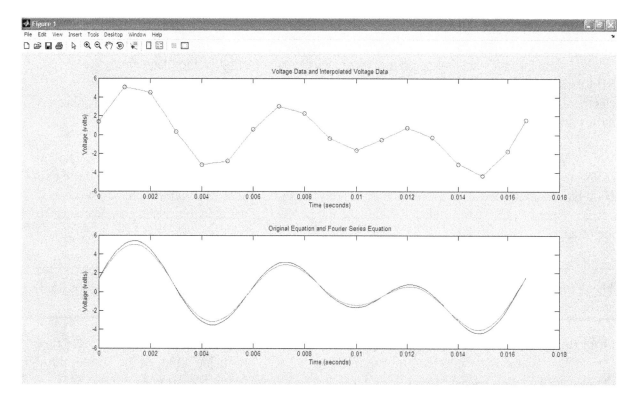

Figure 3-4-4-3 Plot of original data, interpolated data, original equation, and equation created from the computed Fourier Series coefficients.

3.4.5 SIMULINK EXAMPLE

Simulink is a windows-based graphical interface to MATLAB. Simulink stands for SIMulation and LINK. It is used for modeling, simulating, and analyzing dynamic systems. In electrical engineering, Simulink is most often used on control circuits, such as those controlling a motor drive. It is also used in signal processing and communications.

With Simulink, systems are drawn on-screen with functions represented by blocks. The blocks represent transfer functions, summing junctions, etc. Including specialized blocks, more than 1000 blocks are available. Inputs can be on-screen function generators. Outputs can be on-screen oscilloscopes or other display blocks. Simulink and MATLAB are able to transfer data from one to the other.

Simulink can handle simulations that are much more complicated than the one shown here. The demos on the MathWorks website should be viewed by those wanting to learn more. The demos are on http://www.mathworks.com/products/simulink/demos.html?show=demo

Problem:
Use Simulink to redo the problem of Section 3.3.2., producing a plot of the current through the circuit of Figure 3-4-5-1 versus time. The values are repeated: VS = 10 volts, R1 = 3 Ω, L1 = 3 mH, and C1 = 3 microF. The switch closes at time t = 0 seconds, the initial current is 0 amps, and the initial voltage across the capacitor is 0 volts.

Simulink is demonstrated in this example.

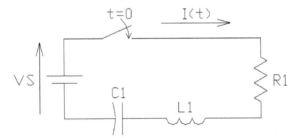

Figure 3-4-5-1 DC capacitor/inductor/resistor circuit.

Solution:
1) To use Simulink the circuit is converted to one with a step voltage. The new circuit diagram is shown in Figure 3-4-5-2.

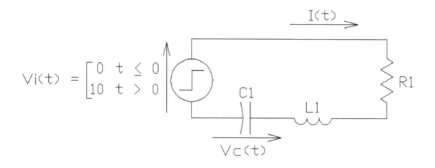

Figure 3-4-5-2 DC capacitor/inductor/resistor circuit modified for analysis by Simulink.

2) As in Section 3.3.2 the loop equation for the circuit is:

$$Vi(t) = R1 \cdot I(t) + L1 \cdot dI(t)/dt + (1/C1) \cdot \int I(t)dt$$

Here Vi represents the combined 10 VDC voltage source and switch as a step function that goes from 0 to 10 volts at time 0^+.

3) The equations are re-written as:

(ii) $\qquad Vi(t) - Vc(t) = R1 \cdot I(t) + L1 \cdot dI(t)/dt$

(iii) $\qquad I(t) = C1 \cdot dVc(t)/dt$

Substituting in I(t)

(iv) $\qquad Vi(t) - Vc(t) = R1 \cdot C1 \cdot dVc(t)/dt + L1 \cdot C1 \cdot dVc(t)^2/dt^2$

Solving for $dVc(t)^2/dt^2$

(v) $\qquad dVc(t)^2/dt^2 = [1/(L1 \cdot C1)] \cdot [Vi(t) - R1 \cdot C1 \cdot dVc(t)/dt - Vc(t)]$

R1, C1, and L1 are inserted to make the numerical equations:

(vi) $\qquad dVc(t)^2/dt^2 = 1.1111 \times 10^8 \cdot Vi(t) - 10^3 \cdot dVc(t)/dt - 1.1111 \times 10^8 \cdot Vc(t)$

(vii) $\qquad I(t) = 3.3333 \times 10^5 \cdot dVc(t)/dt$

4) To start Simulink, first start MATLAB. Then in MATLAB left-click the Simulink icon. It is just below the "MATLAB Desktop" main menu and looks like a red clock with drawing triangles behind it. This starts the "Simulink Library Browser" shown in Figure 3-4-5-3.

5) Open an empty Simulink Model window by left clicking on "File" in the "MATLAB Desktop" main menu and then on "New" and "Model". If necessary, move and reduce in size the windows so that both it and the "Simulink Library Browser" can be seen at the same time. This is shown in Figure 3-4-5-3.

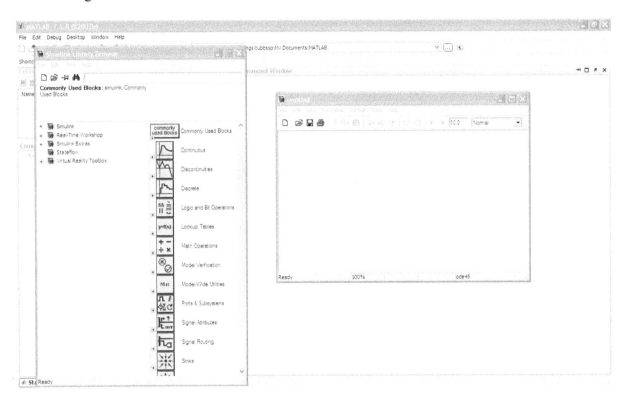

Figure 3-4-5-3 "Simulink Library Browser" and "Model" windows.

6) The numerical equations are represented in Simulink.

a) To create a "Step" source, Vi(t), go to the "Simulink Library Browser", left-click on "Simulink" and "Sources". Then, left-click and drag the "Step" icon onto the "Model" window. Set the icon by releasing the left mouse button. Double left-click on it to bring up the "Source Block Parameters: Step" window. In this window, set the "Step time" at 0, the "Initial value" at 0, and the "Final value" at 10. This is shown in Figure 3-4-5-4.

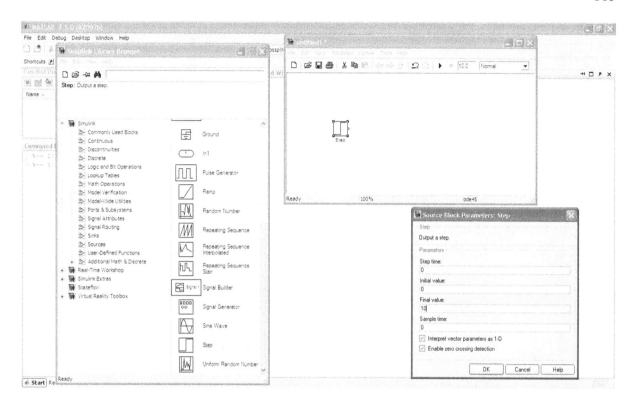

Figure 3-4-5-4 "Simulink Library Browser", "Model", and "Source Block Parameters: Step" windows with step value selected.

b) In the numerical equation (vi), the value of the step voltage, Vi(t), is multiplied by 1.1111×10^8. In the Simulink diagram this is achieved with a 1.1111×10^8 "Gain" block as shown in Figure 3-4-5-5. The "Gain" block is dragged from the "Simulink Library Browser" window's "Commonly Used Blocks". The 1.1111×10^8 value is set by double left-clicking the "Gain" block and typing in 1.1111e8. The 1.1111e8 value appears in the diagram as "K".

c) The output of the "Step" is connected to the input of the "Gain" by left-clicking and holding down the left mouse button on the "Step" and dragging the connection to the input of the "Gain". See Figure 3-4-5-5.

116

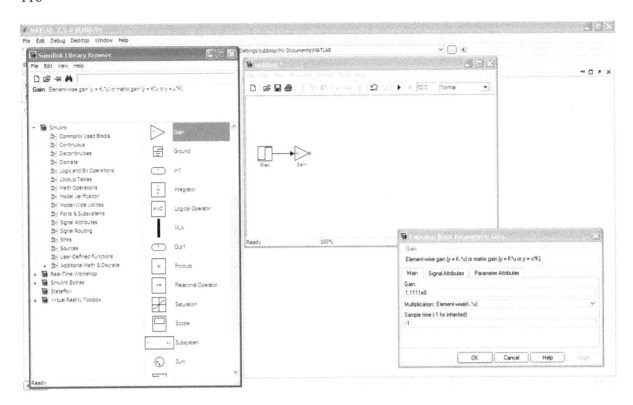

Figure 3-4-5-5 "Simulink Library Browser", "Model", and "Commonly Used Blocks: Gain" window with gain value selected. "Step" is sending its output to the "Gain" input.

d) In the numerical equation $1.1111 \times 10^8 \cdot Vi(t)$ is subtracted by $10^3 \cdot dVc(t)/dt$ and $1.1111 \times 10^8 \cdot Vc(t)$. In Simulink this is done with a "Sum" block. (Note the "Sum" block is not a "Bus" block.) The "Sum" block is configured with one positive input and two negative inputs. It is changed to a rectangular configuration to make it clearer. It can be seen in Figure 3-4-5-6.

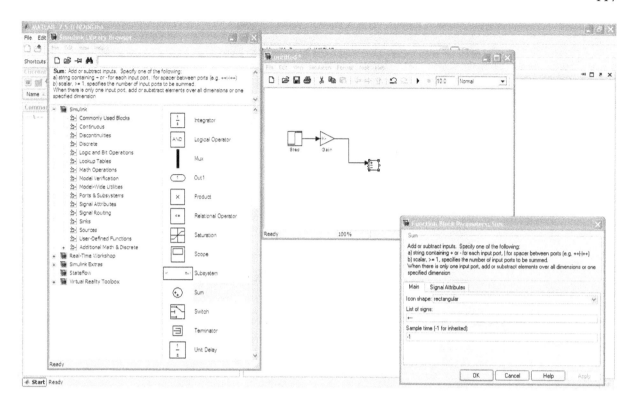

Figure 3-4-5-6 "Simulink Library Browser", "Model", and "Commonly Used Blocks: Sum" window with "Sum" configuration.

e) The output of the "Sum" block is integrated twice. The first integration gives the value of dVc(t)/dt which is then multiplied by a 10³ "Gain" block and then fed back into a "Sum" block input. The second integration gives the value of Vc(t) which is then multiplied by a 1.1111 x 10⁸ "Gain" block and fed back into the other "Sum" block input. This completes the equation. See Figure 3-4-5-7.

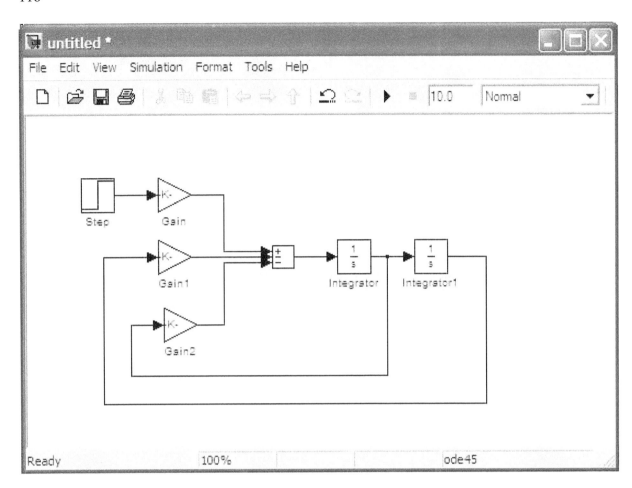

Figure 3-4-5-7 Simulink model for equation (vi).

f) The current through the circuit is determined by the equation I(t) = 3.3333 x
10^5·dVc(t)/dt. A "Gain" and "Derivative" block are added to the Simulink model.

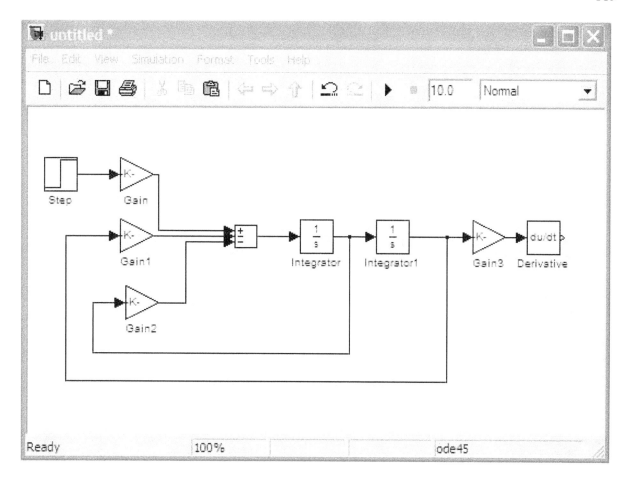

Figure 3-4-5-8 Completed Simulink model for equations (vi) and (vii).

g) "Scope" blocks are added to the model to show the input voltage, the voltage across the capacitor, and the current through the circuit.

120

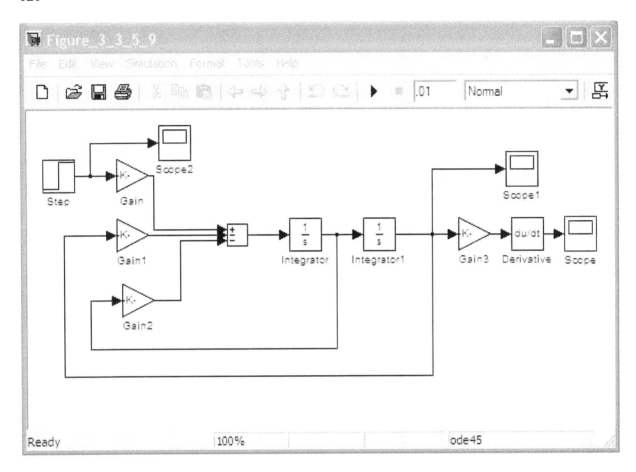

Figure 3-4-5-9 Completed Simulink model for equations (vi) and (vii) with "Scope" outputs for input voltage Vi(t) (Scope2), capacitor voltage Vc(t) (Scope1), and current I(t) (Scope).

      h) Parameters can be set before running Simulink. Left-click on "Simulation" in the "Model" window and then "Configuration Parameters". There are many parameters to choose from. Most of them will be left at default values. Here, only the "Stop Time" is changed from 10 to .005 seconds. See Figure 3-4-5-10.

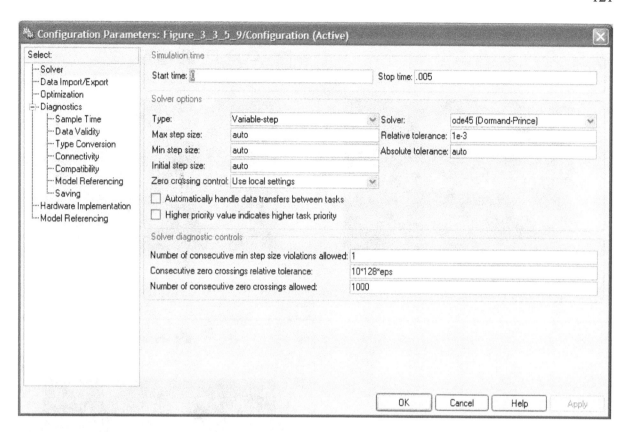

Figure 3-4-5-10 Simulink "Configuration Parameters" "Solver" page with .005 seconds "Stop time" selected.

7) To run the model, left-click on "Simulation" in the "Model" window. Then select "Start". To see the outputs, left-click on the "Scope"'s. "Scope" plots will appear, but will probably have incorrect Y axis ranges. Also, the "Scope" outputs tend to appear on screen on top of each other. In Figure 3-4-5-11 the plots have been dragged to better locations. Also, the Y axes have been properly chosen on each by right-clicking on the plots and selecting "Autoscale".

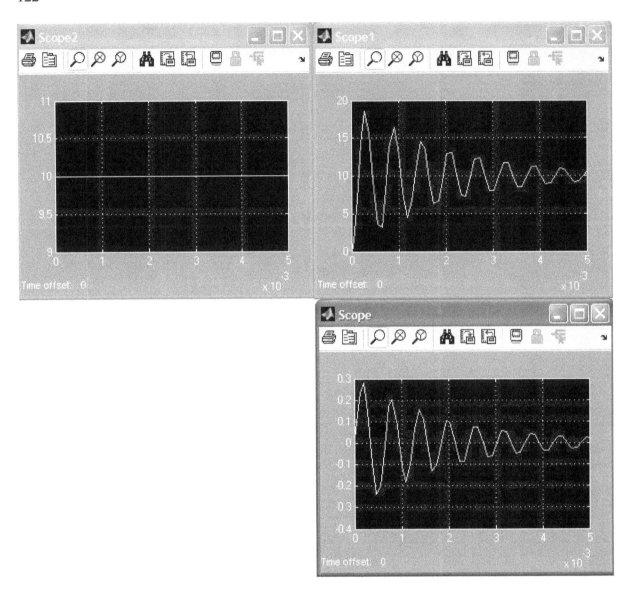

Figure 3-4-5-11 "Stop Time" is set to .005 seconds. Simulink outputs. "Scope2" plot shows the input voltage step, Y axis in volts and X axis in seconds. The "Scope1" plot shows the voltage across the capacitor, Y axis in volts and X axis in seconds. The "Scope" plot shows the current through the circuit, Y axis in amps and X axis in seconds.

8) One of Simulink's limitations can be demonstrated by selecting a longer time of evaluation. With the "Stop Time" reset to 10 seconds, the programs fails to produce a reasonable output. Output plots are shown in Figure 3-4-5-12.

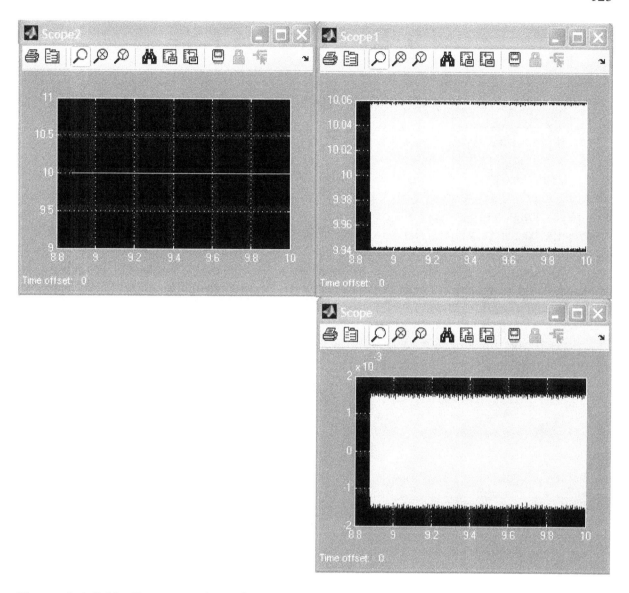

Figure 3-4-5-12 Demonstration of unreasonable outputs when the "Stop Time" is set to 10 seconds. Simulink outputs. "Scope2" plot shows the input voltage step, Y axis in volts and X axis in seconds. The "Scope1" plot shows the voltage across the capacitor, Y axis in volts and X axis in seconds. The "Scope" plot shows the current through the circuit, Y axis in amps and X axis in seconds.

4.0 REFERENCES

1) *MATLAB 7 Getting Started Guide*, The MathWorks. A free copy of it is available on The MathWorks, Inc. website.

It prints on 272 pages. Anybody using MATLAB should get a copy.

2) Pratap, Rudra, *Getting Started with MATLAB 7: A Quick Introduction for Scientists and Engineers*, 2005, Oxford Series in Electrical and Computer Engineering. It is paperback, 275 pages long, and 9.1" x 7.5". Its list price is $35.95. True to its title, this book quickly teaches the basics. It is one of the more popular introductory MATLAB books.

3) Other books on MATLAB. A search on a popular online bookseller brought up 16 books devoted to teaching MATLAB basics and 10 devoted to teaching MATLAB basics to engineers and scientists.

4) Books on Simulink. A search on a popular online bookseller brought up 3 books devoted to teaching Simulink basics and many more that cover both Simulink and MATLAB.

5.0 APPENDIX

5.1 NUMERIC PRECISION

By default, MATLAB stores all numeric values as floating point double-precision (approximately 16 decimal digits). It is possible to specify lesser precision.

By default, MATLAB generally displays 5 or 6 digits. The displayed figures can be adjusted using the "format" statement.

5.2 TIPS

Check MATLAB results for reasonableness. Occasionally MATLAB will produce an incorrect result. One example of this is seen in the Simulink example in Figure 3-3-5-12.

Put comments in programs to tell the future reader what was done.

If you are having trouble writing a program, *get a small part of it running and then build on that.*

To exit out of an operating program, *type "Ctrl c".*

5.3 MATLAB OPERATORS

To get help on any statement or command, block the command in the "Command Window" or MATLAB "Editor" and press F1.

MATLAB has more operators than most electrical engineers need. Below is a listing of the operators demonstrated in this book.

Arithmetic Calculations
    See examples in all sections.

Calculus
    Differential Equations
        See examples in Sections 3.3.1, 3.3.2, 3.3.3, and 3.4.5.
    Differentiation
        See example in Section 3.4.5.
    Integration
        See example in Sections 3.2.10 and 3.4.4.

126

Complex Number (Phasor) Calculations
See examples in Sections 3.2.3, and 3.2.4.

Conditional Statements
See examples in Sections 3.2.5.1, 3.2.8, 3.2.9, and 3.2.10.

Discrete Fourier Transform
See example in Section 3.4.3.

Fitting a Polynomial to Data
See example in Section 3.4.2.

Interpolation
See examples in Sections 3.2.8 and 3.4.4

Matrices and Arrays
See examples in Sections 3.2.2, 3.2.4, 3.2.6, 3.2.7, 3.2.8, 3.2.9, 3.3.3, 3.4.1, 3.4.2, and 3.4.4.

Numerical Precision
See examples in Sections 3.4.2 and 5.1.

Optimization Toolbox
See example in Section 3.2.5.2.

Plotting and Graphing
See examples in Sections 3.2.6. 3.2.7, 3.2.8, 3.2.9, 3.2.10, 3.3.1, 3.3.2, 3.3.3, 3.4.1, 3.4.2, 3.4.3, 3.4.4, and 3.4.5.

Polynomial Approximation
See example in Section 3.4.2.

Program Loops
See examples in Sections 3.2.5.1, 3.2.6, 3.2.8, 3.2.9, 3.2.10, and 3.4.4.

Simulink
See example in Section 3.4.5.

Statistics
See example in Section 3.4.1.

Subroutines
See example in Section 3.2.5.2, 3.3.1, 3.3.2, and 3.3.3.